IFIP Advances in Information and Communication Technology 569

Editor-in-Chief

Kai Rannenberg, Goethe University Frankfurt, Germany

Editorial Board Members

IFIP – The International Federation for Information Processing

IFIP was founded in 1960 under the auspices of UNESCO, following the first World Computer Congress held in Paris the previous year. A federation for societies working in information processing, IFIP's aim is two-fold: to support information processing in the countries of its members and to encourage technology transfer to developing nations. As its mission statement clearly states:

IFIP is the global non-profit federation of societies of ICT professionals that aims at achieving a worldwide professional and socially responsible development and application of information and communication technologies.

IFIP is a non-profit-making organization, run almost solely by 2500 volunteers. It operates through a number of technical committees and working groups, which organize events and publications. IFIP's events range from large international open conferences to working conferences and local seminars.

The flagship event is the IFIP World Computer Congress, at which both invited and contributed papers are presented. Contributed papers are rigorously refereed and the rejection rate is high.

As with the Congress, participation in the open conferences is open to all and papers may be invited or submitted. Again, submitted papers are stringently refereed.

The working conferences are structured differently. They are usually run by a working group and attendance is generally smaller and occasionally by invitation only. Their purpose is to create an atmosphere conducive to innovation and development. Refereeing is also rigorous and papers are subjected to extensive group discussion.

Publications arising from IFIP events vary. The papers presented at the IFIP World Computer Congress and at open conferences are published as conference proceedings, while the results of the working conferences are often published as collections of selected and edited papers.

IFIP distinguishes three types of institutional membership: Country Representative Members, Members at Large, and Associate Members. The type of organization that can apply for membership is a wide variety and includes national or international societies of individual computer scientists/ICT professionals, associations or federations of such societies, government institutions/government related organizations, national or international research institutes or consortia, universities, academies of sciences, companies, national or international associations or federations of companies.

More information about this series at http://www.springer.com/series/6102

Gilbert Peterson · Sujeet Shenoi (Eds.)

Advances in Digital Forensics XV

15th IFIP WG 11.9 International Conference
Orlando, FL, USA, January 28–29, 2019
Revised Selected Papers

 Springer

Editors
Gilbert Peterson
Department of Electrical and Computer
Engineering
Air Force Institute of Technology
Wright-Patterson AFB, OH, USA

Sujeet Shenoi
Tandy School of Computer Science
University of Tulsa
Tulsa, OK, USA

ISSN 1868-4238 ISSN 1868-422X (electronic)
IFIP Advances in Information and Communication Technology
ISBN 978-3-030-28754-2 ISBN 978-3-030-28752-8 (eBook)
https://doi.org/10.1007/978-3-030-28752-8

This Springer imprint is published by the registered company Springer Nature Switzerland AG
The registered company address is: Gewerbestrasse 11, 6330 Cham, Switzerland

Contents

Contents

Contributing Authors

Temilola Aderibigbe recently received his M.S. degree in Computer Science from Florida A&M University, Tallahassee, Florida. His research interests are in the area of digital forensics.

Sudhir Aggarwal is a Professor of Computer Science at Florida State University, Tallahassee, Florida. His research interests include password cracking, mobile forensics, information security and building software systems for digital forensics.

Gunnar Alendal is a Special Investigator with Kripos/NCIS Norway, Oslo, Norway; and a Ph.D. student in Computer Security at the Norwegian University of Science and Technology, Gjovik, Norway. His research interests include digital forensics, reverse engineering, security vulnerabilities, information security and cryptography.

Albert Antwi-Boasiako is the National Cybersecurity Advisor, Republic of Ghana, Ghana, Accra; and the Founder of the e-Crime Bureau, Accra, Ghana. His research interests are in the area of digital forensics, with a focus on digital forensic process standardization.

Stefan Axelsson is an Associate Professor of Digital Forensics at the Norwegian University of Science and Technology, Gjovik, Norway; and an Associate Professor of Digital Forensics at Halmstad University, Halmstad, Sweden. His research interests include digital forensics, data analysis and digital investigations.

Ramesh Babu Battula is an Assistant Professor of Computer Science and Engineering at Malaviya National Institute of Technology, Jaipur, India. His research interests include secure communications, cyber security, performance modeling and next generation networks.

Chun-Fai Chan is a Ph.D. student in Computer Science at the University of Hong Kong, Hong Kong, China. His research interests include penetration testing, digital forensics and Internet of Things security.

Raymond Chan is a Lecturer of Information and Communications Technology at the Singapore Institute of Technology, Singapore. His research interests include cyber security, digital forensics and critical infrastructure protection.

Ramaswamy Chandramouli is a Senior Computer Scientist in the Computer Security Division at the National Institute of Standards and Technology, Gaithersburg, Maryland. His research interests include security for virtualized infrastructures, and smart card interface specification and testing.

Saheb Chhabra is a Ph.D. student in Computer Science and Engineering at Indraprastha Institute of Information Technology, New Delhi, India. His research interests include image processing and computer vision, and their applications to document fraud detection.

Hongmei Chi is an Associate Professor of Computer and Information Sciences at Florida A&M University, Tallahassee, Florida. Her research interests include information assurance, scientific computing, Monte Carlo and quasi Monte Carlo techniques, and data science.

Kam-Pui Chow is an Associate Professor of Computer Science at the University of Hong Kong, Hong Kong, China. His research interests include information security, digital forensics, live system forensics and digital surveillance.

Gokila Dorai is a Ph.D. student in Computer Science at Florida State University, Tallahassee, Florida. Her research interests include computer, mobile device and Internet of Things forensics.

Geir Olav Dyrkolbotn is a Major in the Norwegian Armed Forces, Lillehammer, Norway; and an Associate Professor of Cyber Defense at the Norwegian University of Science and Technology, Gjovik, Norway. His research interest include cyber defense, reverse engineering, malware analysis, side-channel attacks and machine learning.

Struan Gray is an Associate Professor of Physics at Halmstad University, Halmstad, Sweden. His research interests include scanning tunneling microscopy and atomic force microscopy.

Zachary Grimmett is a Computer Engineer with the U.S. Department of Defense in Washington, DC. His research interests include mobile communications devices, digital forensics and malware analysis.

Nicholas Guerra is an M.S. student in Computer Science at the University of Tulsa, Tulsa, Oklahoma. His research interests include digital forensics, cyber security and reverse engineering.

Garima Gupta is a Postdoctoral Researcher in Computer Science and Engineering at Indraprastha Institute of Information Technology, New Delhi, India. Her research interests include image processing and computer vision, and their applications to document fraud detection.

Gaurav Gupta is a Scientist E in the Ministry of Information Technology, New Delhi, India. His research interests include mobile device security, digital forensics, web application security, Internet of Things security and security in emerging technologies.

Monika Gupta is a Visiting Assistant Professor of Optical Physics at Miranda House, Delhi University, India. Her research interests include image processing and computer vision, and their applications to document fraud detection.

Manuel Hernandez is a Software Engineer at Microsoft, Redmond, Washington. His research interests include software engineering and computer hardware.

Yongheng Jia is an M.S. student in Computer Science at Tianjin University, Tianjin, China. His research interests include malware detection and classification.

Umit Karabiyik is an Assistant Professor of Computer and Information Technology at Purdue University, West Lafayette, Indiana. His research interests include digital forensics, user and data privacy, machine learning, and computer and network security.

Martin Karresand is a Senior Scientist at the Swedish Defence Research Agency, Linkoping, Sweden; and a Ph.D. student in Computer Security at the Norwegian University of Science and Technology, Gjovik, Norway. His research interests include digital forensics, file carving, data analysis, machine learning and intrusion detection.

Yuze Li is an M.S. student in Computer Science at Tianjin University, Tianjin, China. His research interests include digital forensics and deep learning.

David Lindahl is a Research Engineer at the Swedish Defence Research Agency, Linkoping, Sweden. His research interests include cyber warfare, critical infrastructure protection and digital forensics.

Changwei Liu is a Postdoctoral Researcher in the Department of Computer Science at George Mason University, Fairfax, Virginia. Her research interests include network security, cloud security and digital forensics.

Jingcheng Liu is an M.S. student in Computer Science at Tianjin University, Tianjin, China. His research interests include data privacy and intrusion detection.

Liangfu Lu is an Assistant Professor of Mathematics at Tianjin University, Tianjin, China. His research interests include compressed sensing, sparse representation and image processing.

Cesar Mak is a Research Programmer at the Logistics and Supply Chain MultiTech R&D Centre, Hong Kong, China. His research interests include digital forensics, machine learning and data analytics.

Tathagata Mukherjee is an Assistant Professor of Computer Science at the University of Alabama in Huntsville, Huntsville, Alabama. His research interests include cyber security, adversarial machine learning, large-scale digital forensics, cyber law, computational geometry, graph theory and optimization.

Martin Olivier is a Professor of Computer Science at the University of Pretoria, Pretoria, South Africa. His research focuses on digital forensics – in particular, the science of digital forensics and database forensics.

James Parsons is a Software Engineer at Microsoft, Redmond, Washington. His research interests include digital forensics and software engineering.

Heloise Pieterse is a Senior Researcher and Software Developer at the Council for Scientific and Industrial Research, Pretoria, South Africa; and a Ph.D. student in Computer Science at the University of Pretoria, Pretoria, South Africa. Her research interests include digital forensics and cyber security.

Emmanuel Pilli is an Associate Professor of Computer Science and Engineering at Malaviya National Institute of Technology, Jaipur, India. His research interests include cyber security, digital forensics, cloud computing, big data, blockchains and the Internet of Things.

Khushboo Rathi is a Senior Software Engineer with Dell Technologies, Round Rock, Texas. Her research interests include digital forensics, mobile forensics and machine learning.

Lakshminarayana Sadineni is a Ph.D. student in Computer Science and Engineering at Malaviya National Institute of Technology, Jaipur, India. His research interests include Internet of Things security and forensics.

Sujeet Shenoi is the F.P. Walter Professor of Computer Science and a Professor of Chemical Engineering at the University of Tulsa, Tulsa, Oklahoma. His research interests include critical infrastructure protection, industrial control systems and digital forensics.

Bhupendra Singh is an Assistant Professor of Computer Science and Engineering at the Indian Institute of Information Technology, Pune, India. His research interests include digital forensics, filesystem analysis and user activity analysis in Windows and Linux systems.

Shweta Singh is an Integrated Software System Engineer at Elkosta Security Systems, New Delhi, India. Her research interests include machine learning and its applications to document fraud detection.

Anoop Singhal is a Senior Computer Scientist and Program Manager in the Computer Security Division at the National Institute of Standards and Technology, Gaithersburg, Maryland. His research interests include network security, network forensics, cloud security and data mining.

Jason Staggs is an Adjunct Assistant Professor of Computer Science at the University of Tulsa, Tulsa, Oklahoma. His research interests include telecommunications networks, industrial control systems, critical infrastructure protection, security engineering and digital forensics.

Renier van Heerden is the Science Engagement Officer at the South African Research and Education Network in Pretoria, South Africa. His research interests include network security, password security and network attacks.

Wynand van Staden is a Senior Lecturer of Computer Science at the University of South Africa, Florida Park, South Africa. His research interests include digital forensics, anonymity and privacy.

Hein Venter is a Professor of Computer Science at the University of Pretoria, Pretoria, South Africa. His research interests are in the area of digital forensics, with a focus on digital forensic process standardization.

Asalena Warnqvist is a Forensics Expert at the National Forensic Centre, Swedish Police Authority, Linkoping, Sweden. Her research interests include digital forensics and data recovery.

Duminda Wijesekera is a Professor of Computer Science at George Mason University, Fairfax, Virginia. His research interests include systems security, digital forensics and transportation systems.

Rodney Wilson is a Software Developer at IBM, Research Triangle Park, North Carolina. His research interests are in the area of software engineering and test automation.

Yaping Zhang is an Assistant Professor of Computer Science at Tianjin University, Tianjin, China. His research interests include network security, data mining and digital forensics.

Preface

Digital forensics deals with the acquisition, preservation, examination, analysis and presentation of electronic evidence. Computer networks, cloud computing, smartphones, embedded devices and the Internet of Things have expanded the role of digital forensics beyond traditional computer crime investigations. Practically every crime now involves some aspect of digital evidence; digital forensics provides the techniques and tools to articulate this evidence in legal proceedings. Digital forensics also has myriad intelligence applications; furthermore, it has a vital role in cyber security – investigations of security breaches yield valuable information that can be used to design more secure and resilient systems.

This book, *Advances in Digital Forensics XV*, is the fifteenth volume in the annual series produced by the IFIP Working Group 11.9 on Digital Forensics, an international community of scientists, engineers and practitioners dedicated to advancing the state of the art of research and practice in digital forensics. The book presents original research results and innovative applications in digital forensics. Also, it highlights some of the major technical and legal issues related to digital evidence and electronic crime investigations.

This volume contains fourteen revised and edited chapters based on papers presented at the Fifteenth IFIP WG 11.9 International Conference on Digital Forensics, held in Orlando, Florida, USA on January 28-29, 2019. The papers were refereed by members of IFIP Working Group 11.9 and other internationally-recognized experts in digital forensics. The post-conference manuscripts submitted by the authors were rewritten to accommodate the suggestions provided by the conference attendees. They were subsequently revised by the editors to produce the final chapters published in this volume.

The chapters are organized into five sections: Forensic Models, Mobile and Embedded Device Forensics, Filesystem Forensics, Image Forensics, and Forensic Techniques. The coverage of topics highlights the richness and vitality of the discipline, and offers promising avenues for future research in digital forensics.

This book is the result of the combined efforts of several individuals. In particular, we thank Mark Pollitt and Jane Pollitt for their tireless work on behalf of IFIP Working Group 11.9. We also acknowledge the support provided by the U.S. National Science Foundation, U.S. National Security Agency and U.S. Secret Service.

GILBERT PETERSON AND SUJEET SHENOI

I

FORENSIC MODELS

Chapter 1

A HOLISTIC FORENSIC MODEL FOR THE INTERNET OF THINGS

Lakshminarayana Sadineni, Emmanuel Pilli and Ramesh Babu Battula

Abstract   The explosive growth of the Internet of Things offers numerous innovative applications such as smart homes, e-healthcare, smart surveillance, smart industries, smart cities and smart grids. However, this has significantly increased the threat of attacks that exploit the vulnerable surfaces of Internet of Things devices. It is, therefore, immensely important to develop security solutions for protecting vulnerable devices and digital forensic models for recovering evidence of suspected attacks. Digital forensic solutions typically target specific application domains such as smart wearables, smart surveillance systems and smart homes. What is needed is a holistic approach that covers the diverse application domains, eliminating the overhead of employing *ad hoc* models.

This chapter presents a holistic forensic model for the Internet of Things that is based on the ISO/IEC 27043 international standard. The model has three phases – forensic readiness (proactive), forensic initialization (incident) and forensic investigation (reactive) – that cover the entire lifecycle of Internet of Things forensics. The holistic model, which provides a customizable and configurable environment that supports diverse Internet of Things applications, can be enhanced to create a comprehensive framework.

Keywords: Internet of Things forensics, holistic forensic model, forensic readiness

1.   Introduction

The Internet of Things (IoT) is a global infrastructure that enables advanced services by interconnecting (physical and virtual) objects based on existing, evolving and interoperable information and communications technologies [14]. The Internet of Things connects electronic, electrical and non-electrical objects to provide seamless communications and contextual services [17]. The explosive growth of Internet of Things devices,

© IFIP International Federation for Information Processing 2019
Published by Springer Nature Switzerland AG 2019
G. Peterson and S. Shenoi (Eds.): Advances in Digital Forensics XV, IFIP AICT 569, pp. 3–18, 2019.
https://doi.org/10.1007/978-3-030-28752-8_1

and the nature of services they provide and data they generate have contributed to an increase in security and privacy breaches as well as other abuses [1, 8]. The need to investigate these incidents has led to the new discipline of Internet of Things forensics, which focuses on the identification, collection, organization and presentation of evidence related to incidents in Internet of Things infrastructures [23].

This chapter presents a holistic forensic model for Internet of Things environments that is based on the ISO/IEC 27043 international standard. The forensic model has three phases, forensic readiness (proactive component), forensic initialization (incident component) and forensic investigation (reactive component). These three phases cover the entire lifecycle of Internet of Things forensics. This chapter also discusses the challenges involved in implementing the forensic model, along with feasible approaches and supporting technologies. The holistic model, which provides a customizable and configurable environment that supports diverse Internet of Things applications, can be enhanced to create a comprehensive framework.

2.   Related Work

The Internet of Things stretches over several layers comprising heterogeneous devices, interconnected networks and diverse communications protocols and applications. Figure 1 shows a typical Internet of Things layered architecture. The three layers – things layer, edge layer and applications layer – are physically and logically divided according to their functionalities. Ideally, the things and edge layers are part of same network and are physically close to each other. As a result, most Internet of Things forensic approaches consider these two layers; the applications layer is left to cloud forensics [15].

Although much research has focused on computer forensics, network forensics and cloud forensics, limited work has been done in the area of Internet of Things forensics. The main reasons are the heterogeneity of devices, and diverse communications protocols and applications domains. These make it very difficult to identify common attack surfaces and create generic security and forensic solutions.

Nevertheless, several researchers have proposed models or frameworks for security analyses and forensic investigations in Internet of Things environments. Oriwoh and Sant [18] have proposed a model for automating security and forensic services that exclusively targets smart home environments. The layered model has four stages. In stage 1, services such as network traffic monitoring, intrusion detection, data collection, parsing, compression and analysis are configured. Stage 2 automates

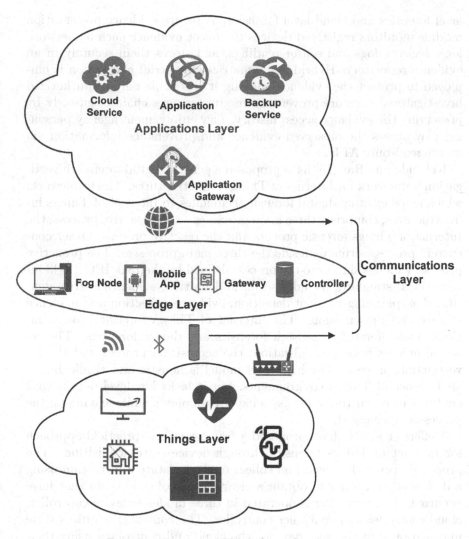

Figure 1. Internet of Things layered architecture.

the configured services to detect incidents and report them to users. In stage 3, users respond to incidents and escalate them to forensic investigators. In stage 4, digital forensic investigators reconstruct the incidents for potential legal action.

Zawoad and Hasan [23] have proposed a forensics-aware model for supporting reliable investigations in Internet of Things environments. Internet of Things forensics has three layers – device level forensics, network level forensics and cloud level forensics. A secure evidence preservation module monitors registered devices to collect evidence such as network logs, registry logs and sensor readings, and stores them securely in an evidence repository. Hybrid (asymmetric-symmetric) encryption is employed to protect the evidence, making it accessible only to authorized investigators. A secure provenance module ensures chain of custody by preserving the evidence access history. Law enforcement agency personnel may access the preserved evidence and provenance information via secure read-only APIs.

Kebande and Ray [15] have proposed a generic digital forensic investigation framework for Internet of Things infrastructures. The framework, which maps existing digital forensic techniques to Internet of Things infrastructures, comprises three main processes – the proactive process, the Internet of Things forensic process and the reactive process. Other concurrent processes run alongside the three main processes. The proactive process, which is similar to the process defined by the ISO/IEC 27043 international standard, includes scenario definition, evidence source identification, planning incident detection, evidence collection and evidence storage and preservation. The Internet of Things forensic process includes cloud forensics, network forensics and device forensics. The reactive process covers initialization, the acquisition process and the investigation process. The high-level model is holistic and applicable to all Internet of Things environments, but it lacks low-level details that enable it to be customized to specific environments while leaving all the processes unchanged.

Meffert et al. [16] have proposed a framework and practical approach for Internet of Things forensics through device state acquisition. The proposed approach is based on collecting device state information using a dedicated controller to obtain a clear picture of the events that have occurred. The controller is operated in three modes – device controller, cloud controller and controller controller. The controller acquires state information directly from devices, the cloud and controllers using their respective modes. While the framework can reliably collect state data in Internet of Things environments using the three modes, its limitations include accessing historical and deleted data, physical access requirements and inability to connect to new devices.

Zia et al. [24] have proposed an application-specific investigative model for Internet of Things environments. The model comprises three independent components – application-specific forensics, digital forensics and forensic process. It is conceptualized based on three key Internet of Things applications – smart homes, wearables and smart cities. The sources of forensic artifacts in the forensic readiness model are smart homes, wearables, smart cities, networks and the cloud.

Shin et al. [20] focus on the reactive process that occurs after an incident has occurred. They applied various digital forensic methods to collect data from an Internet of Things device (Amazon Echo) and network (home area network using the Z-Wave protocol). However, their approach is limited to selected devices and communications protocols.

Babun et al. [5] have proposed a digital forensic framework for smart environments such as smart homes and smart offices, where applications installed on smart devices are used to control sensors and actuators in the environments. The framework has two components – modifier and analyzer. The modifier examines the source code of smart applications at compile time to detect forensically-relevant data and insert tracing logs in the appropriate places. The analyzer uses data processing and machine learning techniques to extract forensic data related to device activity in the event of an incident.

Harbawi and Varol [11] have proposed an improved digital evidence acquisition model for Internet of Things forensics. They highlight the need to identify things of interest that produce initial evidence traces. Perumal et al. [19] have proposed a four-tiered digital forensic investigation model for the Internet of Things. Their model covers the entire investigative lifecycle starting from the authorization of forensic experts in a case to the archival of evidence after the case is closed.

Unfortunately, the forensic models discussed above fail to provide low-level details on how they can be customized to specific application scenarios. In contrast, the model proposed in this chapter engages a holistic approach that emphasizes configurable forensic readiness that is applicable to any Internet of Things domain.

3.   Proposed Holistic Forensic Model

The proposed holistic forensic model for the Internet of Things is based on the ISO/IEC 27043 international standard [13]. The standard describes digital forensics as comprising several processes, each incorporating one or more activities. ISO/IEC 27043 processes correspond to phases in the proposed model and activities correspond to modules.

Figure 2 presents the holistic forensic model. The model has three phases: (i) forensic readiness (proactive) phase; (ii) forensic initialization (incident) phase; and (iii) forensic investigation (reactive) phase. Each phase has a number of component modules. Although all the modules focus on Internet of Things devices, their approaches can be mapped to the applications layer if needed.

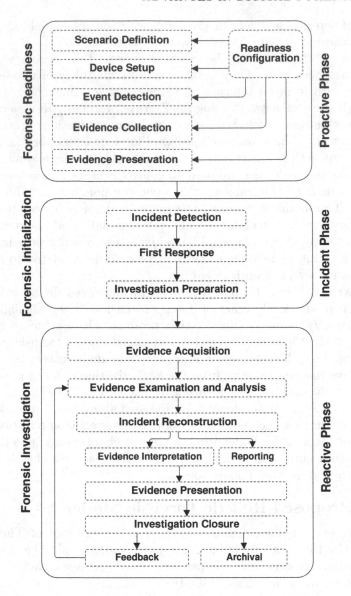

Figure 2.   Holistic forensic model for the Internet of Things.

3.1   Forensic Readiness (Proactive) Phase

During the forensic readiness (proactive) phase, digital evidence related to an Internet of Things environment is collected and preserved.

This reduces the time, effort and cost involved in investigating subsequent incidents.

The forensic readiness phase has six modules.

- **Module 1.1 (Readiness Configuration):** This module coordinates all the forensic readiness activities. It provides configurable services to customize the model to different Internet of Things environments, rendering the model holistic. The configuration is performed by administrators and/or security experts to create application-specific, device-specific and context-aware directions for event detection, forensic data collection and preservation.

  The readiness configuration module has the following basic functionality:

  - Provides a mechanism for adding comprehensive information about Internet of Things devices in an environment (e.g., adding information about the smart devices in a smart home). The information about each device includes the device name, device manufacturer, device type, device id, firmware details, device functionality, interactions to be logged (based on defined scenarios) and device description.
  - Guides the device setup module in identifying suitable properties and configuring each device for evidence collection.
  - Guides the event detection module in identifying the specific events that must be logged.
  - Guides the evidence collection module on the data pertaining to specific events that needs to be collected.
  - Guides the evidence preservation module on how the collected data is formatted and stored for future investigations.

  It is important to note the difference between an event and an incident. An event denotes one or more interactions with Internet of Things devices that can change their states (e.g., changes in the sensor readings of a smart watch and a sensor data request sent from a mobile application to a smart watch); an event need not be suspicious. In contrast, an incident is a sequence of suspicious events that disrupts the regular functioning of Internet devices; an incident impacts security and/or privacy.

- **Module 1.2 (Scenario Definition):** This module defines scenarios as sequences of events that are forensically sensitive to specific Internet of Things applications (e.g., unusual interactions with a

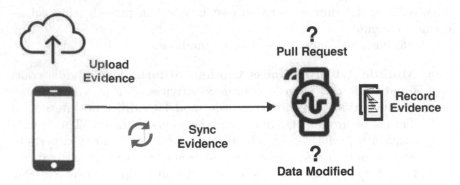

Figure 3. Example event in a smart watch scenario.

device and failed authorization attempts when accessing a service). In the applications layer, scenarios are defined to cover how configuration and business data should be managed (e.g., who can access or modify the data). Each scenario specifies events that change the states of Internet of Things devices. The state changes are identified along with the properties of the associated devices.

- **Module 1.3 (Device Setup):** This module identifies each new device added to the environment and its forensic properties before the device becomes operational. It consults the readiness configuration module for device-specific settings and stores all the setup information in a secure database for use by other modules. Also, it keeps track of when a device is detached from the environment.

- **Module 1.4 (Event Detection):** This module identifies forensically-sensitive events based on scenarios defined in the scenario definition module. Rules may be specified for validating device interactions and network traffic, and identifying potential events. In the applications layer, an autonomous module may be designed to monitor the security aspects of system configurations and requests for authentication and data access.

Figure 3 presents an example event in a smart watch scenario. When sensor data is updated or a pull request is sent from a mobile application to the smart watch, the associated data is recorded as potential evidence. The data is periodically synchronized with a mobile application and may be uploaded to cloud storage for subsequent processing.

- **Module 1.5 (Evidence Collection):** This module covers the collection of potential evidence from Internet of Things devices, controllers and network devices. An example is logging the commands issued to an Internet of Things device along with their timestamps. The sources of commands to the devices are recorded (including multiple possible sources for a device such as a smart TV – TV remote, direct push button and remote user over a network).

  Evidence collection is easier when a device operating system supports forensic interactions to collect relevant information via system calls. Otherwise, an autonomous software layer on top of the operating system has to be created for evidence collection from programmable devices. In both cases, an edge controller issues commands to the device software for evidence collection and preservation. For all other devices, an external collection mechanism is implemented at the controller node.

  When devices execute real-time applications, it is important to know the kind of data that is generated and how it is stored; this helps develop advanced data collection mechanisms [21]. In the applications layer, the sources of all failed interactions (e.g., configuration changes, authentication and access requests, and suspicious API calls) are logged for future investigations. All the collected evidence is formatted according to the storage and processing requirements.

- **Module 1.6 (Evidence Preservation):** This module covers the secure storage of evidence for future investigations. Many Internet of Things devices have on-board flash memory that stores the operating system and real-time executable files. This memory can be used to store forensic data, which could be sent periodically to a central server for longer-term storage and subsequent processing. Alternative storage may be provided by fog nodes. The evidence should be stored securely and protected from accidental modification and intentional tampering. Potential evidence from the applications layer may be preserved in secure cloud storage.

3.2   Forensic Initialization (Incident) Phase

The forensic initialization (incident) phase has three modules: (i) incident detection; (ii) first response; and (iii) investigation preparation.

- **Module 2.1 (Incident Detection):** This module covers the continuous monitoring of an environment for harmful behavior using

appropriate techniques and tools. All user interactions are validated against rules defined by administrators or security experts.
In the device and controller levels, the rules are implemented as
intelligent scripts that identify malicious interactions (e.g., a script
would detect a number of failed authentication requests by an Internet of Things device that exceed a threshold). In the network
level, intrusion detection systems and other security tools are used
to monitor live traffic. In the applications level, cloud security
techniques and tools are used to detect incidents. After an incident is detected, it is reported for further action.

- **Module 2.2 (First Response):** This module covers the transmission of prioritized alerts to users or administrators for immediate action. In the case of an incident, an alert is escalated to a
  digital forensic professional. If required, devices, controllers and
  software are suspended to prevent additional damage to the environment. All the relevant components should be disconnected
  from the production environment until the forensic investigation
  is completed.

- **Module 2.3 (Investigation Preparation):** This module covers
  activities that support the investigative process. The activities
  include:

    - An incident management (investigative) plan is prepared. The
      plan specifies how to proceed with an investigation. It also
      covers evidence provenance and formatting.

    - An incident response team of available experts is created to
      implement the incident management plan. Each incident is
      investigated by a dedicated team.

    - Technical and other support, including organizational and op-
      erational support, are provided to the team.

    - The incident response team is briefed about and trained on
      incident management.

    - The incident management plan is reviewed and improved us-
      ing techniques such as paper tests, tabletop exercises and sim-
      ulations.

    - The improved incident management plan is documented for
      practical implementation.

3.3   Forensic Investigation (Reactive) Phase

The forensic investigation (reactive) phase implements the investigative plan to reconstruct the sequence of events. Potential evidence collected during the readiness phase is acquired and analyzed to prove or disprove that an attack or breach occurred and to identify the victim devices. The insights gained during the investigation are used to improve security and forensic techniques and tools used in the environment.

The forensic investigation phase comprises the following five modules:

- **Module 3.1 (Evidence Acquisition):** This module covers the identification of evidence pertaining to the reported incident and its acquisition from secure storage. It may be necessary to visit the physical location and acquire forensic images of the Internet of Things devices in question. Various techniques may be used to extract the firmware and memory images in order to identify malicious behavior. In the applications layer, artifacts related to the cloud environment such as virtual machine images and logs, hypervisor logs, user activity logs, database access logs and application logs are collected.

- **Module 3.2 (Evidence Examination and Analysis):** This module covers the formatting of the acquired logs and evidence to render them suitable for analysis. Machine learning techniques may be applied to identify attack patterns in Internet of Things networks. Techniques and tools must be updated or augmented periodically in order to identify new attacks. Analytic tools may be used in the applications layer to identify suspicious behavior related to computing, storage and data access requests.

- **Module 3.3 (Incident Reconstruction):** This module covers the reconstruction of an incident as a sequence of suspicious events based on the results of the evidence examination and analysis module. The incident reconstruction module comprises the following two activities:

  - The evidence interpretation activity analyzes results based on predefined postulates to reconstruct an incident (e.g., identify the sequences of events in the devices and edge layer and map them to the applications layer to understand what has occurred). The postulates may be adapted from standard security policies or defined by security experts for specific Internet of Things application scenarios (e.g., a security policy may be defined to limit the number of unsuccessful authentication or

access requests). Some policies may define the standard behavior of Internet of Things devices or the environment to
avoid unwanted communications between extraneous devices.

– The reporting activity generates a formal report covering the
incident findings related to attacks and victims and their
timelines.

■ **Module 3.4 (Evidence Presentation):** This module covers the
preparation and presentation of evidence to comply with the requirements imposed by legal proceedings. The final report may
incorporate graphics and animations to enhance clarity.

■ **Module 3.5 (Investigation Closure):** This module covers the
post-investigation activities, especially providing feedback and ar-
chiving the evidence. Feedback is provided to the evidence examination and analysis module, and evidence traces and records
are archived. Case studies may be created to inform and enhance
future investigations.

4.   Forensic Technologies

This section discusses two emerging technologies, fog/edge computing
and blockchains, that can enhance Internet of Things forensic processes.

4.1   Fog/Edge Computing

The terms fog computing and edge computing are used interchange-
ably to describe the layer between end-devices and the cloud that leverages the storage and processing of intermediate devices (fog nodes). Fog
computing can be considered to be an implementation of edge computing [9]. Edge computing brings down services from the cloud to the
edges of Internet of Things networks. Since these services include device
authentication, access control, and data processing and storage, most
of the forensic readiness modules can be implemented using fog computing. Al-Masri et al. [3] have proposed a fog-based digital forensic
investigation framework for Internet of Things environments.

4.2   Blockchains

The distributed and immutable characteristics of blockchains suit the
demands of Internet of Things forensics. Fernandez-Carames and Fraga-
Lamas [10] have presented a comprehensive decision model that checks
whether or not a blockchain-based solution applies to a particular Internet of Things scenario. In the decision model, evidence collected from

Internet of Things devices, controllers and applications in the cloud are treated as the ledger. An ideal solution for Internet of Things forensics is a private-permissioned blockchain where the number of nodes is restricted and access is only provided to selected users.

The distributed nature of a blockchain dovetails with fog computing to provide services such as evidence collection and storage. Evidence can be collected by any node and updated in the ledger. The immutability of a blockchain ensures that the evidence is not tampered with and is always valid. A blockchain also supports the verification of the provenance of evidence. These two properties enable forensic investigators to access evidence reliably from any node at any time. Ali et al. [2] have presented a global naming and storage system secured by blockchains.

In summary, blockchains can be used to timestamp and store evidence collected from Internet of Things devices [10]. Banerjee et al. [6] have presented an interesting blockchain application that tracks changes made to Internet of Things device firmware and automatically restores the original firmware in the event of tampering. Similar approaches can be used to maintain the integrity of Internet of Things evidence.

5. Research Challenges

Internet of Things forensics is challenging due to the complexity of devices and applications, and the lack of uniform standards across device manufacturers and system developers. Most tools are designed to work with conventional systems with significant storage and computing capabilities instead of small, specialized devices [21]. Challenges are also imposed by the heterogeneity of devices, applications and communications technologies. As a result, the stored data has diverse formats and requires custom acquisition methods.

Another challenge is extracting volatile data from Internet of Things devices before it is overwritten. Sophisticated mechanisms are needed for swift collection. Collection can be sped up by storing data on the device itself, but the data must be moved periodically to supplementary storage to free up device memory. The data may also be synchronized to fog nodes or cloud storage at regular intervals. This approach is safer in the long term because Internet of Things devices can be tampered with or even destroyed. The transfer and aggregation of evidence also make it more difficult to maintain the chain of custody [12]; fortunately, this can be addressed using blockchain technology.

Some challenges are specific to the phases of the proposed holistic forensic model. The principal challenge in the readiness phase is applying forensic processes to devices and their firmware when the devices are operating. Separate hardware devices with automated forensic scripts may have to be developed to support forensic readiness activities. Challenges in the incident phase include taking control of devices deployed at remote locations (software-defined networking could help) and communicating alerts about incidents. Challenges during the investigation phase include formatting heterogeneous evidence into a uniform structure for examination and analysis, and employing machine learning algorithms to detect new attacks (e.g., cross-layer attacks) [4].

6.   Conclusions

Due to the diversity of devices, networks and applications, a number of *ad hoc* digital forensic solutions have been developed for specific Internet of Things environments. A holistic digital forensic model that covers diverse Internet of Things environments is required to eliminate the overhead imposed by the *ad hoc* solutions.

The Internet of Things forensic model presented in this chapter is holistic and covers the entire forensic lifecycle. The model, which is based on the ISO/IEC 27043 international standard, is customizable and configurable, and supports diverse Internet of Things applications.

Future research will focus on the implementation and testing of the model in selected application domains, with the ultimate goal of creating a comprehensive framework for Internet of Things forensics.

References

[1] F. Alaba, M. Othman, I. Hashem and F. Alotaibi, Internet of Things security: A survey, *Journal of Network and Computer Applications*, vol. 88, pp. 10–28, 2017.

[2] M. Ali, J. Nelson, R. Shea and M. Freedman, Blockstack: A global naming and storage system secured by blockchains, *Proceedings of the USENIX Annual Technical Conference*, pp. 181–194, 2016.

[3] E. Al-Masri, Y. Bai and J. Li, A fog-based digital forensics investigation framework for IoT systems, *Proceedings of the Third IEEE International Conference on Smart Cloud*, pp. 196–201, 2018.

[4] V. Asati, E. Pilli, S. Vipparthi, S. Garg, S. Singhal and S. Pancholi, RMDD: Cross-layer attack in Internet of Things, *Proceedings of the International Conference on Advances in Computing, Communications and Informatics*, pp. 172–178, 2018.

[5] L. Babun, A. Sikder, A. Acar and A. Uluagac, IoTDots: A Digital Forensics Framework for Smart Environments, arXiv:1809.00745 (arxiv.org/abs/1809.00745), 2018.

[6] M. Banerjee, J. Lee and K. Choo, A blockchain future for Internet of Things security: A position paper, *Digital Communications and Networks*, vol. 4(3), pp. 149–160, 2018.

[7] M. Chernyshev, S. Zeadally, Z. Baig and A. Woodward, Internet of Things forensics: The need, process models and open issues, *IT Professional*, vol. 20(3), pp. 40–49, 2018.

[8] M. Conti, A. Dehghantanha, K. Franke and S. Watson, Internet of Things security and forensics: Challenges and opportunities, *Future Generation Computer Systems*, vol. 78(2), pp. 544–546, 2018.

[9] K. Dolui and S. Datta, Comparison of edge computing implementations: Fog computing, cloudlet and mobile edge computing, *Proceedings of the Global Internet of Things Summit*, 2017.

[10] T. Fernandez-Carames and P. Fraga-Lamas, A review of the use of blockchain for the Internet of Things, *IEEE Access*, vol. 6, pp. 32979–33001, 2018.

[11] M. Harbawi and A. Varol, An improved digital evidence acquisition model for Internet of Things forensics I: A theoretical framework, *Proceedings of the Fifth International Symposium on Digital Forensics and Security*, 2017.

[12] R. Hegarty, D. Lamb and A. Attwood, Digital evidence challenges in the Internet of Things, *Proceedings of the Ninth International Workshop on Digital Forensics and Incident Analysis*, pp. 163–172, 2014.

[13] International Organization for Standardization and International Telecommunication Union, ISO/IEC 27043:2015: Information Technology – Security Techniques – Incident Investigation Principles and Processes, Geneva, Switzerland, 2015.

[14] International Telecommunication Union, Recommendation ITU-T Y.2060: Overview of the Internet of Things, Geneva, Switzerland, 2012.

[15] V. Kebande and I. Ray, A generic digital forensic investigation framework for Internet of Things (IoT), *Proceedings of the Fourth IEEE International Conference on Future Internet of Things and Cloud*, pp. 356–362, 2016.

[16] C. Meffert, D. Clark, I. Baggili and F. Breitinger, Forensic state acquisition from Internet of Things (FSAIoT): A general framework and practical approach for IoT forensics through IoT device state acquisition, *Proceedings of the Twelfth International Conference on Availability, Reliability and Security*, article no. 65, 2017.

[17] R. Minerva, A. Biru and D. Rotondi, Towards a Definition of the Internet of Things (IoT), Revision 1, IEEE Internet Initiative, Piscataway, New Jersey (iot.ieee.org/images/files/pdf/IEEE_IoT_Towards_Definition_Internet_of_Things_Revision1_27MAY15.pdf), 2015.

[18] E. Oriwoh and P. Sant, The Forensics Edge Management System: A concept and design, *Proceedings of the Tenth IEEE International Conference on Ubiquitous Intelligence and Computing and the Tenth IEEE International Conference on Autonomic and Trusted Computing*, pp. 544–550, 2013.

[19] S. Perumal, N. Norwawi and V. Raman, Internet of Things (IoT) digital forensic investigation model: Top-down forensic approach methodology, *Proceedings of the Fifth International Conference on Digital Information Processing and Communications*, pp. 19–23, 2015.

[20] C. Shin, P. Chandok, R. Liu, S. Nielson and T. Leschke, Potential forensic analysis of IoT data: An overview of the state-of-the-art and future possibilities, *Proceedings of the IEEE International Conference on the Internet of Things, IEEE Green Computing and Communications, IEEE Cyber, Physical and Social Computing and IEEE Smart Data*, pp. 705–710, 2017.

[21] S. Watson and A. Dehghantanha, Digital forensics: The missing piece of the Internet of Things promise, *Computer Fraud and Security*, vol. 2016(6), pp. 5–8, 2016.

[22] K. Yeow, A. Gani, R. Ahmad, J. Rodrigues and K. Ko, Decentralized consensus for edge-centric Internet of Things: A review, taxonomy and research issues, *IEEE Access*, vol. 6, pp. 1513–1524, 2017.

[23] S. Zawoad and R. Hasan, FAIoT: Towards building a forensics aware ecosystem for the Internet of Things, *Proceedings of the IEEE International Conference on Services Computing*, pp. 279–284, 2015.

[24] T. Zia, P. Liu and W. Han, Application-specific digital forensics investigative model in Internet of Things (IoT), *Proceedings of the Twelfth International Conference on Availability, Reliability and Security*, article no. 55, 2017.

Chapter 2

IMPLEMENTING THE HARMONIZED MODEL FOR DIGITAL EVIDENCE ADMISSIBILITY ASSESSMENT

Albert Antwi-Boasiako and Hein Venter

Abstract   Standardization of digital forensics has become an important focus area for researchers and criminal justice practitioners. Over the past decade, several efforts have been made to encapsulate digital forensic processes and activities in harmonized frameworks for incident investigations. A harmonized model for digital evidence admissibility assessment has been proposed for integrating the technical and legal determinants of digital evidence admissibility, thereby providing a techno-legal foundation for assessing digital evidence admissibility in judicial proceedings.

This chapter presents an algorithm underlying the harmonized model for digital evidence admissibility assessment, which enables the determination of the evidential weight of digital evidence using factor analysis. The algorithm is designed to be used by judges to determine evidence admissibility in criminal proceedings. However, it should also be useful to investigators, prosecutors and defense lawyers for evaluating potential digital evidence before it is presented in court.

Keywords: Digital evidence admissibility, factor analysis, evidential weight

1.   Introduction

The application of digital forensics in criminal justice has become more relevant than ever because of the continuous evolution of cyber crime and its impact on individuals, organizations and governments. It is nearly impossible in today's information-technology-driven society to find a crime that does not have a digital dimension [7]. The relevance of digital forensics is also influenced by the fact that computer systems are being used to facilitate crimes such as fraud, terrorism and money laundering. National information infrastructures have become targets

© IFIP International Federation for Information Processing 2019
Published by Springer Nature Switzerland AG 2019
G. Peterson and S. Shenoi (Eds.): Advances in Digital Forensics XV, IFIP AICT 569, pp. 19–36, 2019.
https://doi.org/10.1007/978-3-030-28752-8_2

for cyber attackers; this has rendered digital forensics an essential component of national strategies for combating cyber threats.

Meanwhile, advancements in computer engineering and information and communications technologies have led to novel sources of digital evidence. Unmanned aerial vehicles, driverless automobiles and Internet of Things devices have led to new developments in digital forensics because of the digital evidence that resides in these systems [1, 9].

However, the question of digital evidence admissibility remains a key issue when applying digital forensics in jurisprudence. The criminal justice sector is confronted with the challenge of proffering evidence that is admissible in court [12]. In addition to training in new legislation and technology, judges require a scientific approach for assessing digital evidence in court. These challenges have driven the research community to develop standardized processes and approaches to ensure that digital evidence is admissible in legal proceedings.

This chapter presents an algorithm underlying a harmonized model for digital evidence admissibility assessment, which assists in determining the evidential weight of digital evidence using factor analysis. The algorithm is designed to be used by judges in criminal proceedings, but it should also be useful to investigators, prosecutors and defense lawyers for evaluating potential digital evidence before it is presented in legal proceedings.

2.   Background and Related Work

Several models and frameworks have been introduced to standardize digital forensic activities in order to address issues regarding the admissibility of digital evidence. These include a framework introduced by participants in the 2001 Digital Forensic Research Workshop [17], an abstract model of digital forensic procedure introduced by Reith et al. [18] and a harmonized process model introduced by Valjarevic and Venter [25]. A good practice guide produced by the (U.K.) Association of Chief Police Officers [3] and an electronic crime scene investigation guide published by the U.S. Department of Justice [23] are examples of efforts undertaken by law enforcement to harmonize digital forensics and provide a common approach for conducting digital investigations. The International Organization for Standardization has created the ISO/IEC 27037 Standard [13] and the ISO/IEC 27043 Standard [14] to support incident investigations.

Despite significant developments in rationalizing the domain of digital forensics, issues associated with the admissibility of digital evidence in legal proceedings have remained largely unresolved. To address this

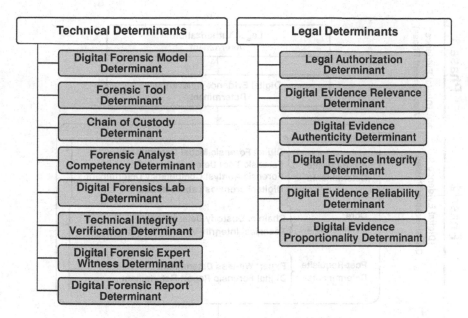

Figure 1. Requirements for assessing the admissibility of digital evidence.

gap, Antwi-Boasiako and Venter [2] introduced the Harmonized Model for Digital Evidence Admissibility Assessment (HM-DEAA). This model specifies technical and legal requirements – called "determinants" – that underpin the admissibility of digital evidence. Figure 1 presents the various technical and legal determinants specified in the harmonized model.

This existential foundation of digital evidence presents a techno-legal dilemma – a challenge or gap that exists in establishing a balanced interdependent relationship between the technical and legal requirements when establishing digital evidence admissibility and determining the weight of digital evidence in judicial proceedings. The harmonized model of Antwi-Boasiako and Venter [2] leverages an operational interdependency relationship between the technical and legal determinants to establish digital evidence admissibility.

Figure 2 presents the harmonized model. The three phases of the model are integrated, but they are distinct from each other due to their functional relevance in assessing digital evidence admissibility. The digital evidence assessment phase establishes the legal foundations of digital evidence. The digital evidence consideration phase focuses on the technical requirements that underpin digital evidence admissibility. The digital

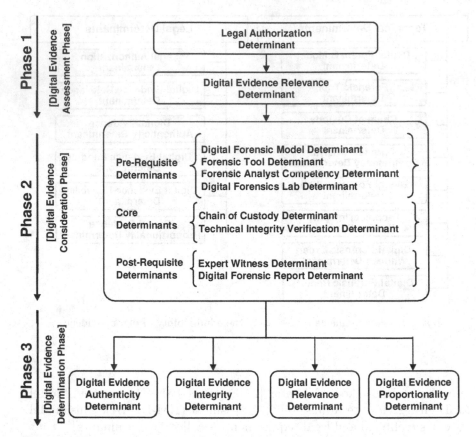

Figure 2. Harmonized Model for Digital Evidence Admissibility Assessment.

evidence determination phase underpins the judicial decisions regarding the admissibility and weight of digital evidence.

The research described in this chapter builds on the previous work by Antwi-Boasiako and Venter [2]. It presents an algorithm that underlies the implementation of the harmonized model for digital evidence admissibility assessment and enables the determination of evidential weight using factor analysis.

3.   Validation Survey Methodology and Findings

A survey of judicial experts with knowledge and experience in digital evidence was conducted to validate the technical and legal determinants of digital evidence admissibility. The respondents were asked to assess

Table 1. Evidential weight impact description.

Score	Impact	Description
1	No Impact	Determinant has no effect on the digital evidence in question
2	Minimal	Determinant has very little effect on the digital evidence in question
3	Moderate	Determinant has some effect, but not significant enough, on the digital evidence in question
4	Significant	Determinant has considerable effect on the digital evidence in question
5	Very Significant	Determinant has exceptional effect on the digital evidence in question

the impact of each determinant on the weight of digital evidence. Table 1 shows the Likert scale [4] used by the survey respondents.

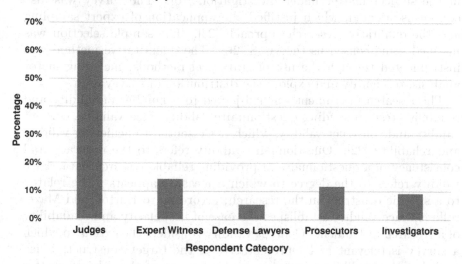

Figure 3. Survey respondent categories.

A total of 77 respondents participated in the survey. The respondents were drawn from common law and civil law jurisdictions across Africa, North and South America, Asia, Europe and the Middle East. Figure 3 shows the five categories of experts who participated in the survey.

An expert sampling method [10] used to obtain a scientifically-valid sample for the survey. Expert sampling provides an optimal means for constructing the views of respondents who are judged to be experts

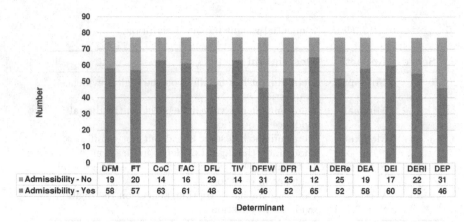

	DFM	FT	CoC	FAC	DFL	TIV	DFEW	DFR	LA	DERe	DEA	DEI	DERI	DEP
■ Admissibility - No	19	20	14	16	29	14	31	25	12	25	19	17	22	31
■ Admissibility - Yes	58	57	63	61	48	63	46	52	65	52	58	60	55	46

Determinant

Figure 4. Responses related to the determinants of admissibility.

in the subject matter under investigation [10]. The survey was also consensus-oriented, which justified the application of expert sampling and the qualitative research approach [24]. The sample selection was justified using consensus theory [8, 26]. The quantitative method was instrumented through the use of statistical methods, including factor analysis, to identify and explore the distribution of survey data.

The research instrument was subjected to a number of validity and reliability tests, including questionnaire validity, face validity, content validity and construct validity, which are essential to achieving validity and reliability [22]. Questionnaire validity refers to the accuracy and consistency of a questionnaire in providing reliable research data. Face validity refers to the degree to which a measure appears to be related to a specific construct in the research; according to Burton and Mazerolle [6], face validity establishes the ease of use, clarity and readability of a research instrument. Content validity considers the extent to which a survey is relevant and representative of the target construct; it establishes the credibility, accuracy and relevance of the subject matter under investigation. Construct validity establishes a cause and effect relationship in a research instrument [22].

Figure 4 highlights the responses related to the determinants of admissibility. As an example, consider the chain of custody (CoC) determinant. Fourteen survey participants (18% of the respondents) indicated that chain of custody does not affect the admissibility of digital evidence in a court of law whereas 62 participants (82% of the respondents) indicated that it affects evidence admissibility. Several factors may have contributed to these responses. Chain of custody is widely recognized by

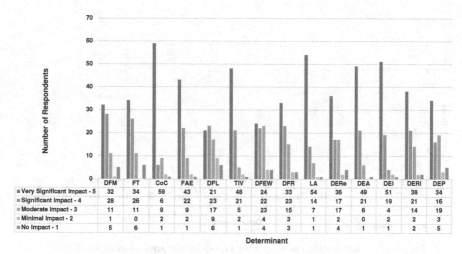

	DFM	FT	CoC	FAE	DFL	TIV	DFEW	DFR	LA	DERe	DEA	DEI	DERI	DEP
▪ Very Significant Impact - 5	32	34	59	43	21	48	24	33	54	36	49	51	38	34
▪ Significant Impact - 4	28	26	6	22	23	21	22	23	14	17	21	19	21	16
▪ Moderate Impact - 3	11	11	9	9	17	5	23	15	7	17	6	4	14	19
▪ Minimal Impact - 2	1	0	2	2	9	2	4	3	1	2	0	2	2	3
▪ No Impact - 1	5	6	1	1	6	1	4	3	1	4	1	1	2	5

Determinant

Figure 5. Likert scores assigned to the determinants of admissibility.

experts as one of the most important requirements for digital evidence admissibility; this is confirmed by the high positive response rate of 82% for the determinant. However, the understanding of respondents and prevailing legal practices in their jurisdictions may have contributed to the higher than expected 18% negative response rate for chain of custody.

The survey participants were also asked to rate the impact of each determinant on the evidential weight using the Likert scale of 1 to 5 shown in Table 1. Figure 5 shows the scores for the determinants. Once again, consider the chain of custody determinant (CoC) as an example. Fifty-nine survey participants (77% of the total) rated the impact of chain of custody on digital evidence admissibility as very significant (Likert score of 5); six respondents (8%) rated the impact as significant (score of 4); nine 9 respondents (12%) rated the impact as moderate (score of 3); two respondents (3%) rated the impact as minimal (score of 2); and one respondent (less than 1%) rated no impact (score of 1).

Figure 6 graphs the minimum, average and maximum scores for the determinants. For example, the average rating of the impact of the chain of custody determinant on digital evidence admissibility is 4.53. It is important to note that an analysis of the data revealed that no conspicuous variations existed in the responses provided by judges versus other criminal justice actors relative to the importance of the determinants. This implies that all the criminal justice actors considered in the research have common understanding and expectations of the application of dig-

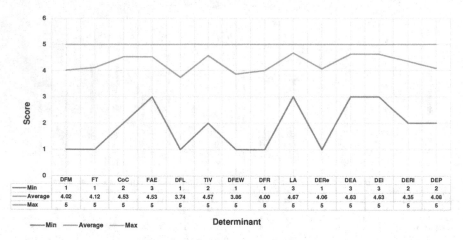

	DFM	FT	CoC	FAE	DFL	TIV	DFEW	DFR	LA	DERe	DEA	DEI	DERI	DEP
Min	1	1	2	3	1	2	1	1	3	1	3	3	2	2
Average	4.02	4.12	4.53	4.53	3.74	4.57	3.86	4.00	4.67	4.06	4.63	4.63	4.35	4.08
Max	5	5	5	5	5	5	5	5	5	5	5	5	5	5

——Min ——Average ——Max   **Determinant**

Figure 6.   Distributions of scores for the determinants of admissibility.

ital evidence in criminal proceedings. However, the levels of technical
and judicial knowledge and experience appear to be important factors
that contributed to the variations seen in the scores.

4.   Proposed Algorithm

The next step after validating the determinants and assessing their
impacts on digital evidence admissibility is to apply the algorithm presented in Figures 7 and 8. The algorithm flowcharts cover the three
phases of the harmonized model: (i) digital evidence assessment; (ii)
digital evidence consideration; and (iii) digital evidence determination.
The algorithm formalizes the sequential activities from the introduction
of digital evidence in court through the various stages of witness presentation and cross-examination to the final determination of the case by
the court.

During the first phase, digital evidence assessment, the legal foundations of digital evidence are established. The relevance of the evidence
to the case is determined by the court after legal authorization is established. This phase covers pre-trial activities in most jurisdictions. The
trial could be terminated at this stage if a proper legal foundation is not
established.

If the proper legal foundation is established, the case moves to full
trial corresponding to the second phase – digital evidence consideration. The prerequisite requirements, core requirements and evaluation
requirements, which are all technical determinants listed in Figure 2, are
assessed during this phase.

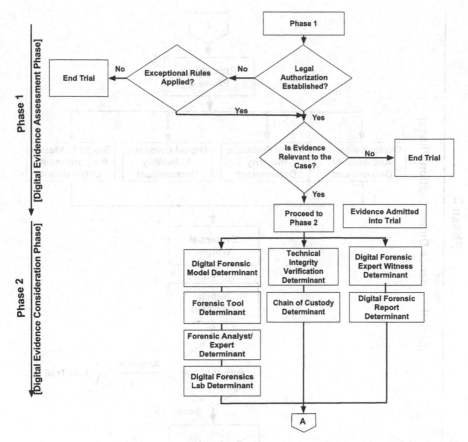

Figure 7. Flowchart of the digital evidence assessment and consideration phases.

The third phase, digital evidence determination, forms the basis of judicial decisions. In most jurisdictions, the decision could be acquittal or conviction and sentencing. The sentence would be the maximum, average or minimum based on the evidential weight established through the operationalization of the harmonized model.

5. Evidential Weight Determination

This section presents the foundation for determining the evidential weight of digital evidence using the determinants discussed in this chapter.

Evidential weight is the weight that a judge would attach to a particular piece of evidence that is tendered in a court of law. According to Mason [15], assessing evidential weight involves scrutinizing a piece

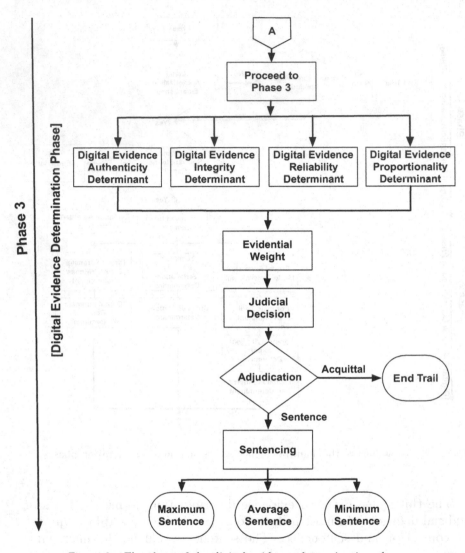

Figure 8. Flowchart of the digital evidence determination phase.

of evidence and deciding whether or not it is acceptable and relevant to arriving at a decision during a trial.

The research described in this chapter employed factor analysis [5] to statistically analyze the survey data in order to determine evidential weight. Factor analysis was selected because it is well suited to exploratory data analyses. In particular, it was used to obtain the weights of the variables required to make judicial decisions. The survey conducted in this research provided the data used to operationalize factor analysis [16].

In order for a dataset to be suitable for factor analysis, a correlation must exist between the determinants and it must pass the Kaiser-Meyer-Olkin (KMO) sampling adequacy test. The correlations between the determinants were computed using the sample Pearson correlation coefficient [21] as follows:

$$r = \frac{N \sum xy - (\sum x)(\sum y)}{\sqrt{[N \sum x^2 - (\sum x)^2][N \sum y^2 - (\sum y)^2]}} \tag{1}$$

where r is the correlation coefficient between determinants x and y (x and y are the individual survey responses); N is the number of survey respondents; $\sum xy$ is the sum of the products of paired x and y scores; $\sum x$ is the sum of x scores; $\sum y$ the sum of y scores; $\sum x^2$ the sum of squared x scores; and $\sum y^2$ is the sum of squared y scores.

Note that the correlation is calculated for each pair of determinants. Also, the numerator in the equation is the covariance between the two determinants and the denominator is the product of the standard deviations of the two determinants.

The Stata statistical software package [19] was used to compute the correlations. For example, a correlation of 0.324962 was established between the forensic tool (FT) and digital forensic model (DFM) determinants, and a correlation of 0.500934 was established between the legal authorization (LA) and technical integrity verification (TIV) determinants.

The KMO sampling adequacy test was performed to ensure that the dataset was suitable for factor analysis. The KMO sampling adequacy varies from zero to one; a value close to one denotes well suited to factor analysis whereas a value close to zero denotes inappropriate for factor analysis. A KMO sampling adequacy value of 0.77 was obtained, suggesting that the dataset is adequate for factor analysis [20].

Factor analysis assumes that a linear relationship involving the latent factors exists in the survey data. In general, a factor $factor_{nj}$ in the data is expressed as:

$$factor_{nj} = b_1 X_{1j} + b_2 X_{2j} + ... + b_n X_{nj} + e_j \tag{2}$$

where the b_i terms denote factor loadings (e.g., factor scores such as that relating determinant FT to determinant DFM as computed by Stata); X_{ij} terms correspond to the determinants; j is an observation (i.e., factor); n is the number of variables (i.e., number of determinants); and e_j is an error term.

The coefficient formula for the determinants is given by:

$$
\begin{aligned}
\text{Factor Analysis of Determinants} \;=\; & b_1 DFM + b_2 FT + b_3 CoC + \\
& b_4 FAC + b_5 DFL + b_6 TIV + \\
& b_7 DFEW + b_8 DFR + b_9 LA + \\
& b_{10} DERe + b_{11} DEA + \\
& b_{12} DEI + b_{13} DERI + \\
& b_{14} DEP + e_j \qquad (3)
\end{aligned}
$$

The b_i values in Equation (3) are used to compute the evidential weight EW as follows:

$$
\begin{aligned}
EW \;=\; & w_1 DFM + w_2 FT + w_3 CoC + \\
& w_4 FAC + w_5 DFL + w_6 TIV + \\
& w_7 DFEW + w_8 DFR + \\
& w_9 LA + w_{10} DERe + \\
& w_{11} DEA + w_{12} DEI + \\
& w_{13} DERI + w_{14} DEP + \\
& e_j \qquad (4)
\end{aligned}
$$

where the w_i terms correspond to the determinant weights Wd_i computed as:

$$
Wd_i = \frac{b_i n^2}{\text{Total Variance}} \qquad (5)
$$

Note that i denotes a determinant; n is the number of determinants; b_i is a factor score generated by factor analysis; and the total variance is the sum of the squares of the b_i factor scores.

Table 2 presents the computed factor loadings $b_i n^2$ and determinant weights Wd_i based on the survey results.

6.   Results and Discussion

The equations presented in the previous section were applied to a hypothetical case involving digital evidence. Table 3 presents the results obtained by applying factor analysis to evidence in the hypothetical case. In the table, a determinant weight Wd_i denotes the weight of determinant i as established by factor analysis. A determinant score Sd_i in the table, which corresponds to the score assigned to determinant i by the court for the case in question, is given by:

Table 2. Evidential weight determination.

Determinant	Factor Loading (b_i)	Factor Score $(b_i n^2)$	Determinant Weight (Wd_i)
DFM	0.247633	0.061322	0.034
FT	0.412889	0.170477	0.095
CoC	0.344163	0.118448	0.066
FAC	0.372313	0.138617	0.025
DFL	0.212455	0.045137	0.077
TIV	0.371712	0.138170	0.077
DEFW	0.237606	0.056457	0.031
DFR	0.326640	0.106694	0.059
LA	0.240957	0.058060	0.032
DERe	0.193218	0.037333	0.021
DEA	0.495371	0.245393	0.136
DEI	0.611801	0.374300	0.208
DERl	0.332325	0.110440	0.061
DEP	0.375614	0.141086	0.078
Total Variance		1.801933	

Table 3. Evidential weight determination and analysis.

Determinant	Determinant Weight (Wd_i)	Determinant Score (Sd_i)	Weighted Value (Wv_i)
DFM	0.034	3.8	0.129
FT	0.095	4.5	0.428
CoC	0.066	3.0	0.198
FAC	0.025	2.5	0.063
DFL	0.077	3.4	0.262
TIV	0.077	2.3	0.177
DFEW	0.031	5.0	0.155
DFR	0.059	4.7	0.277
LA	0.032	3.7	0.118
DERe	0.021	4.2	0.088
DEA	0.136	4.0	0.544
DEI	0.208	2.4	0.499
DERl	0.061	3.6	0.220
DEP	0.078	3.5	0.273
Total Evidential Weight	3.431		

$$Sd_i = \frac{\text{Sum of Assessment Scores}}{\text{Total Mark}} \times 5 \qquad (6)$$

where each determinant has a maximum mark allocation of five.

Each of determinants is assessed in court using different parameters, which are essentially the key questions addressed during evidence presentation and cross-examination. For example, relative to the digital forensic tool (FT) determinant, the following key questions are considered to determine the score:

- Which forensic tool(s) was/were used in the forensic examination?

- Was the use of each tool licensed?

- Was open-source or proprietary software used?

- What are the implications of using each tool?

- Was each tool tested or validated?

- What is the error rate of each tool?

- What is the level of acceptance of each tool by the researcher and practitioner communities?

- Are there any scientific publications about each tool?

The answers to these questions are determined based on scientific and industry requirements in order to accept a forensic tool in digital investigations. While the questions are not exhaustive, they provide key assessment parameters that would be used in court to provide a score for the given determinant. A score of 4.5 for the digital forensic tool determinant was obtained by applying Equation 6. This value was computed for the determinant based on the assessment questions.

Using Equation 4 and the data in Table 3, the evidential weight is computed as:

$$
\begin{aligned}
EW = \ & 0.034DFM + 0.095FT + \\
& 0.066CoC + 0.025FAE + \\
& 0.077DFL + 0.077TIV + \\
& 0.031DFEW + 0.059DFR + \\
& 0.032LA + 0.021DERe + \\
& 0.136DEA + 0.208DEI + \\
& 0.061DERI + 0.078DEP
\end{aligned}
\qquad (7)
$$

The weighted value Wv_i, which corresponds to the evidential weight of determinant i, is computed as:

$$Wv_i = Wd_i \times Sd_i \qquad (8)$$

where Wd_i is the weight of determinant i and Sd_i is the determinant score.

Thus, the total weighted value of all the determinants is given by:

$$\sum_{i=1}^{n} Wd_i Sd_i = Wd_1 Sd_1 + Wd_2 Sd_2 + Wd_3 Sd_3 + \ldots + Wd_n Sd_n \qquad (9)$$

where n is the number of determinants.

Upon inserting the values from Table 3, the value of the evidential weight is computed as:

$$
\begin{aligned}
EW &= (0.034 \times 3.8) + (0.095 \times 4.5) + (0.066 \times 3) + \ldots (0.078 \times 3.5) \\
&= 3.431 \qquad (10)
\end{aligned}
$$

Expressing the evidential weight as a percentage $EW\%$ yields:

$$
\begin{aligned}
EW\% &= \frac{EW}{5} \times 100 \\
&= \frac{3.431}{5} \times 100 \\
&= 68.62 \qquad (11)
\end{aligned}
$$

The evidential weight of 3.431, which corresponds to 68.62%, is tendered in court and provides the basis for a judicial decision. The percentage value of the evidential weight could guide the court on the sentencing level, which can be the maximum, average or minimum sentence. However, it should be noted that judicial decisions are also impacted by other mitigating factors. This is because judges have certain discretionary powers under the law that they may exercise when they deem necessary. The mitigating factors include the age of the accused, guilty plea, number of years already spent in custody, demonstration of remorse and other extenuating factors.

While there are limits to applying the harmonized model in judicial proceedings, it is important to emphasize that mitigating factors are considered after the model has provided a judge with scientific guidance to make a judicial decision. Therefore, any mitigating factors and the discretionary powers given to a judge as an arbiter of justice do not affect the scientificness of the harmonized model as a judicial tool.

7.   Conclusions

The algorithm presented in this chapter operationalizes the harmonic model for digital evidence admissibility assessment and customizes the model to enable the determination of evidential weight. The algorithm and evidential weight determination are designed to be used by judges in criminal proceedings. They should also be useful to investigators, prosecutors and defense lawyers for evaluating potential digital evidence before it is presented in legal proceedings.

It is important to note that advances in digital forensics are expected to impact the results of future surveys of the type conducted in this research. Different results in future surveys would result in different weights to the determinants as well as different sets of determinants. Such changes are to be expected in the rapidly-evolving field of digital forensics. Nevertheless, the harmonized model, survey research methodology and evidential weight determination framework are sound and robust, implying that surveys would have to be conducted periodically to generate new data, determinants and determinant weights that will keep up with trends in digital forensics and how digital evidence is used in legal proceedings.

Future research will focus on developing an expert system that operationalizes the harmonic model for digital evidence admissibility assessment. The expert system, which will draw on concepts from computational forensics [11], could be applied in real cases, including jury trials, to establish the utility of the harmonized model across the various types of criminal proceedings.

References

[1] S. Alabdulsalam, K. Schaefer, T. Kechadi and N. Le-Khac, Internet of Things forensics: Challenges and a case study, in *Advances in Digital Forensics XIV*, G. Peterson and S. Shenoi (Eds.), Springer, Cham, Switzerland, pp. 35–48, 2018.

[2] A. Antwi-Boasiako and H. Venter, A model for digital evidence assessment, in *Advances in Digital Forensics XIII*, G. Peterson and S. Shenoi (Eds.), Springer, Cham, Switzerland, pp. 23–38, 2017.

[3] Association of Chief Police Officers, Good Practice Guide for Computer-Based Evidence, London, United Kingdom, 2008.

[4] D. Bertram, Likert Scales ...are the Meaning of Life, CPSC 681 – Topic Report (`poincare.matf.bg.ac.rs/~kristina/topic-dane-likert.pdf`), 2008.

[5] A. Bryman and D. Cramer, Constructing variables, in *Handbook of Data Analysis*, M. Hardy and A. Bryman (Eds.), SAGE Publications, London, United Kingdom, pp. 18–34, 2004.

[6] L. Burton and S. Mazerolle, Survey instrument validity, Part I: Principles of survey instrument development and validation in athletic training education research, *Athletic Training Education Journal*, vol. 6(1), pp. 27–35, 2011.

[7] E. Casey, *Digital Evidence and Computer Crime: Forensic Science, Computers and the Internet*, Academic Press, Waltham, Massachusetts, 2011.

[8] D. Child, *The Essentials of Factor Analysis*, Bloomsbury Academic, London, United Kingdom, 2006.

[9] T. Cowper and B. Levin, Autonomous vehicles: How will they challenge law enforcement? *Law Enforcement Bulletin*, FBI Training Division, Federal Bureau of Investigation, Quantico, Virginia (leb.fbi.gov/articles/featured-articles/autono mous-vehicles-how-will-they-challenge-law-enforcement), February 13, 2018.

[10] I. Etikan and K. Bala, Sampling and sampling methods, *Biometrics and Biostatistics International Journal*, vol. 5(6), article no. 00148, 2017.

[11] K. Franke and S. Srihari, Computational forensics: An overview, *Proceedings of the Second International Workshop on Computational Forensics*, pp. 1–10, 2008.

[12] S. Goodison, R. Davis and B. Jackson, Digital Evidence and the U.S. Criminal Justice System: Identifying Technology and Other Needs to More Effectively Acquire and Utilize Digital Evidence, Technical Report RR 890-NIJ, RAND Corporation, Santa Monica, California, 2015

[13] International Organization for Standardization, Information Technology – Security Techniques – Guidelines for Identification, Collection, Acquisition and Preservation of Digital Evidence, ISO/IEC 27037:2012 Standard, Geneva, Switzerland, 2012.

[14] International Organization for Standardization, Information Technology – Security Techniques – Incident Investigation Principles and Processes, ISO/IEC 27043:2015 Standard, Geneva, Switzerland, 2015.

[15] S. Mason, *Electronic Evidence*, Butterworths Law, London, United Kingdom, 2012.

[16] Organisation for Economic Co-operation and Development, *Handbook on Constructing Composite Indicators: Methodology and User Guide*, OECD Publishing, Paris, France, 2008.

[17] G. Palmer, A Road Map for Digital Forensic Research, DFRWS Technical Report, DTR-T001-01 Final, Air Force Research Laboratory, Rome, New York, 2001.

[18] M. Reith, C. Carr and G. Gunsch, An examination of digital forensic models, *International Journal of Digital Evidence*, vol. 1(3), 2002.

[19] StataCorp, Stata Release 15, College Station, Texas (`www.stata.com/products`), 2019.

[20] Statistics How To, Kaiser-Meyer-Olkin (KMO) Test for Sampling Adequacy (`statisticshowto.com/kaiser-meyer-olkin`), 2016.

[21] Study.com, Pearson Correlation Coefficient: Formula, Example and Significance, Mountain View, California (`study.com/academy/lesson/pearson-correlation-coefficient-formula-example-significance.html`), 2019.

[22] H. Taherdoost, Validity and reliability of the research instrument: How to test the validation of a questionnaire/survey in a research, *International Journal of Academic Research in Management*, vol. 5(3), pp. 28–36, 2016.

[23] Technical Working Group for Electronic Crime Scene Investigation, Electronic Crime Scene Investigation: A Guide for First Responders, NIJ Guide, NCJ 187736, U.S. Department of Justice, Washington, DC, 2001.

[24] R. Trotter, Qualitative research sample design and sample size: Resolving and unresolved issues and inferential imperatives, *Preventive Medicine Journal*, vol. 55(5), pp. 398–400, 2012.

[25] A. Valjarevic and H. Venter, Harmonized digital forensic process model, *Proceedings of the Information Security for South Africa Conference*, 2012.

[26] S. Weller and A. Romney, *Systematic Data Collection*, SAGE Publications, Newbury Park, California, 1988.

II

MOBILE AND EMBEDDED DEVICE FORENSICS

Chapter 3

CLASSIFYING THE AUTHENTICITY OF EVALUATED SMARTPHONE DATA

Heloise Pieterse, Martin Olivier and Renier van Heerden

Abstract   Advances in smartphone technology coupled with the widespread use of smartphones in daily activities create large quantities of smartphone data. This data becomes increasingly important when smartphones are linked to civil or criminal investigations. As with all forms of digital data, smartphone data is susceptible to intentional or accidental alterations by users or installed applications. It is, therefore, essential to establish the authenticity of smartphone data before submitting it as evidence. Previous research has formulated a smartphone data evaluation model, which provides a methodical approach for evaluating the authenticity of smartphone data. However, the smartphone data evaluation model only stipulates how to evaluate smartphone data without providing a formal outcome about the authenticity of the data.

    This chapter proposes a new classification model that provides a grade of authenticity for evaluated smartphone data along with a measure of the completeness of the evaluation. Experimental results confirm the effectiveness of the proposed model in classifying the authenticity of smartphone data.

Keywords: Mobile device forensics, smartphone data, authenticity

1.   Introduction

The competitive nature of the global smartphone market [4] stimulates continuous advancements in smartphone technology. The advancements enable smartphone models to support different operating systems and permit the installation of diverse third-party applications. The current capabilities of smartphones coupled with their widespread use in daily activities lead to rich collections of data. Smartphone data "includes any data of probative value that is generated by an application or transferred to the smartphone by the end-user" [12]. Generally, smartphone data

© IFIP International Federation for Information Processing 2019
Published by Springer Nature Switzerland AG 2019
G. Peterson and S. Shenoi (Eds.): Advances in Digital Forensics XV, IFIP AICT 569, pp. 39–57, 2019.
https://doi.org/10.1007/978-3-030-28752-8_3

describes events that occurred on the smartphone and the associated timestamps support the chronological ordering of the events [1]. As a result, smartphone data constitutes valuable digital evidence in civil and criminal investigations.

Smartphone data is, however, susceptible to modification [7]. Changes to smartphone data can occur during the execution of incorrect or error-prone applications or deployed malware. Furthermore, users with malicious intent can alter smartphone data intentionally. Intentional changes to smartphone data are commonly referred to as anti-forensics, which "compromise[s] the availability or usefulness of evidence to the forensic process" [8]. While several studies have successfully demonstrated the manipulation, fabrication and alteration of smartphone data [11, 14], unknown or unexpected changes to smartphone data that go undetected can lead to erroneous conclusions in investigations. Therefore, it is essential for digital forensic professionals to establish the authenticity of smartphone data before formulating any conclusions [15]. Authenticity refers to the preservation of data from the time it was first generated and the ability to prove that the integrity of the data has been maintained over time [3, 5, 6, 10].

Establishing the authenticity of smartphone data requires a good understanding of the smartphone operating environment and the key components that are responsible for creating smartphone data. These components include the smartphone applications that generate data, operation of the smartphone by the end-user and the impact of the immediate surroundings.

Pieterse et al. [13] formally defined the term "authenticity" with regard to smartphone data and used the definition to articulate several requirements for evaluating the authenticity of the data. These requirements were subsequently employed to construct a smartphone data evaluation model that provides digital forensic professionals with a structured approach for evaluating the authenticity of smartphone data. However, the data evaluation model only stipulates how to evaluate smartphone data – it does not provide a formal classification of the authenticity of the evaluated data. Meanwhile, classification scales for digital evidence, such as Casey's certainty scale or degrees of likelihood (almost definitely, most probably, probably, very possible or possibly) [3], have been proposed for specifying the certainty of conclusions. A formal and consistent methodology for classifying the authenticity of smartphone data would provide further support to the certainty of investigative conclusions.

This chapter introduces a new classification model for smartphone data, which is constructed using the smartphone data evaluation model and the requirements for evaluating the authenticity of smartphone data. The classification model assesses smartphone data using an ordered pair of values. The first value corresponds to a grade of authenticity while the second value describes the completeness of the evaluation. This classification enables digital forensic professionals to present the authenticity of evaluated smartphone data with confidence. Experiments involving the manipulation of iPhone 7 data confirm the effectiveness of the classification model in assessing the authenticity of smartphone data.

2. Background

A detailed analysis of smartphone data offers contextual information about the end-user as well as the activities performed with the smartphone. Therefore, smartphone data can constitute valuable digital evidence in civil and criminal investigations. The authenticity of smartphone data is of great importance to ensuring that digital forensic professionals draw correct and accurate conclusions based on the data. In order to formulate proper conclusions, digital forensic professionals must be able to review smartphone data and to evaluate its authenticity.

The smartphone data evaluation model of Pieterse et al. [13] offers a methodical approach for evaluating smartphone data. This section briefly reviews the formal definition of authentic smartphone data, the requirements for identifying smartphone data and the smartphone data evaluation model.

2.1 Authentic Smartphone Data

Smartphones operate in interconnected environments where several components are responsible for creating smartphone data. These components are:

- **End-User Behavior:** End-user operation of and interactions with a smartphone.

- **Smartphone Operation:** The working and operational states of a smartphone.

- **Smartphone Application Behavior:** The behavior and execution of installed applications on a smartphone.

- **External Environment:** The roles of mobile networks as delivery platforms.

Authentic smartphone data requires the four components to consistently operate as expected and to remain unaffected. The importance of these components renders them critical to maintaining data authenticity. An affected component that operates irregularly directly impacts data authenticity because an opportunity exists for the data to change. Digital forensic professionals must evaluate all the components in order to establish the authenticity of smartphone data.

2.2   Requirements for Authentic Data

A set of requirements is needed to confirm that the four components operate as expected. The requirements should capture the expected operational behavior of each component, enabling digital forensic professionals to assess the components. The outcomes produced by the requirements would offer digital forensic professionals insights into the authenticity of smartphone data.

Pieterse et al. [12] were the first researchers to identify requirements for evaluating smartphone data. They presented seven theories of normality that capture the normal or expected behavior of smartphone applications. Subsequent research [13] extended the theories of normality by including additional requirements that assess the operation of smartphones and the impacts of the environments external to the smartphones. The remainder of this section discusses the final requirements identified for authentic smartphone data.

The first component covers the end-user and his/her use of the smartphone. Therefore, the requirements evaluate the expected operation of the smartphone and the installed applications as operated by the end-user. The requirements related to the first component are: (1.1) assessing smartphone application usage; (1.2) assessing the operation of the smartphone with regard to rebooting; and (1.3) assessing the presence of anti-forensic applications.

The second component covers the operational state of the smartphone, which reflects the changes made to the smartphone by the end-user. The requirements are: (2.1) assessing the smartphone state (i.e., whether or not the smartphone is rooted or jailbroken); and (2.2) assessing the essence of known critical files. A critical file is one that is used by a digital forensic professional to establish the authenticity of smartphone data.

The third component covers the behavior of the installed smartphone applications. One requirement related to smartphone application behavior is: (3.1) confirming that the internally-stored data corresponds to the data displayed on the user interface (because the data shown on the user interface could be cached data). Another requirement is: (3.2) confirming that the structure (i.e., database) responsible for storing persistent data follows a consistent pattern in storing data (i.e., records are correctly ordered when listed using an auto-incremented primary key and a date or timestamp). In addition: (3.3) confirming that all changes to the file structure (file sizes) occur consistently. An example is a SQLite database that appends new records in a write-ahead log (WAL), which causes the file size to increase. The last requirement is: (3.4) confirming that the ownership and file permissions assigned to the file structure remains unchanged.

The fourth component covers the environment external to the end-user and smartphone. The external environment includes smartphone data collected by other smartphones that directly communicated with the smartphone under investigation, as well as the records collected by mobile network operators. Therefore, the requirements for this component are: (4.1) confirming that the persistent smartphone data stored on two or more smartphones corresponds to the viewed data; and (4.2) confirming that the persistent smartphone data corresponds to the records collected by mobile network operators.

The requirements collectively enable comprehensive reviews of smartphone data as well as the components responsible for creating the data. The outcomes produced by the requirements describe the authenticity of the data and confirm whether or not opportunities existed for the data to be modified. However, the requirements need to be ordered in a formal manner to ensure their optimal use by digital forensic professionals.

2.3   Smartphone Data Evaluation Model

The requirements discussed above provide digital forensic professionals with a mechanism for evaluating smartphone data. However, the absence of structure or order to these requirements can impact their use in investigations. Consequently, the proposed smartphone data evaluation model structures the requirements to provide digital forensic professionals with a step-by-step guide for evaluating and reviewing smartphone data.

The smartphone data evaluation model has three phases: (i) pre-evaluation phase; (ii) evaluation phase; and (iii) documentation phase.

- **Pre-Evaluation Phase:** In this phase, a digital forensic professional performs an inspection of the smartphone. Figure 1 presents the steps involved in this phase. The results produced by the phase describe the smartphone accessibility (i.e., locked or unlocked) and current smartphone state (i.e., rooted or jailbroken), along with the most appropriate data acquisition technique (i.e., logical or physical). Logical acquisition retrieves a bit-for-bit copy of the logical

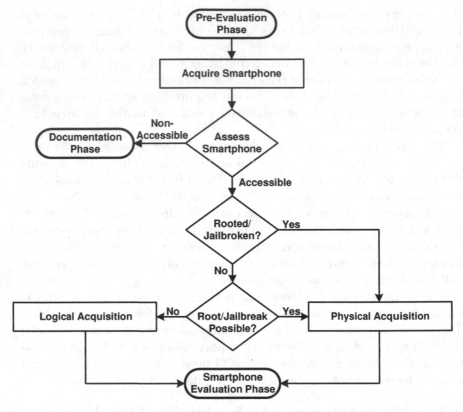

Figure 1. Pre-evaluation phase.

file allocation storage area (filesystem partition), which includes directories and files of various types [2, 9]. Physical acquisition obtains a bit-for-bit copy of the entire physical store (raw disk image), which includes deleted and lost data [2, 9].

- **Evaluation Phase:** The evaluation phase, which follows the pre-evaluation phase, engages the requirements identified in Section 2.2 to review the acquired smartphone data. Figure 2 shows the steps involved in the evaluation phase, which are structured according to the four components identified in Section 2.1.

In the first step of the evaluation phase, a digital forensic professional selects a single smartphone application to be evaluated; this application must reside on the smartphone. After the application is selected, the digital forensic professional interprets and evaluates the collected smartphone data against the requirements of each of

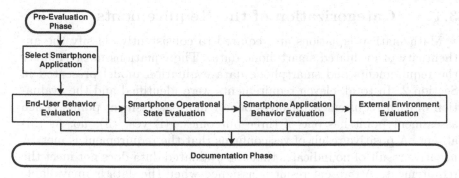

Figure 2. Evaluation phase.

the four components. The outcome of the evaluation phase is a collection of results that offers guidance to the digital forensic professional about the authenticity of the evaluated smartphone data.

- **Documentation Phase:** The final documentation phase of the smartphone data evaluation model involves the collection and aggregation of all the results produced during the evaluation phase. The results enable a digital forensic professional to make informed decisions pertaining to the evaluated smartphone data.

3. Classification Model

The smartphone data evaluation model only stipulates how the data is to be evaluated without providing an outcome regarding the authenticity of the data. Further assistance can be provided to a digital forensic professional by formulating a classification model that assesses the authenticity of the evaluated smartphone data. Collectively, the requirements and smartphone data evaluation model presented in Section 2 provide a foundation for establishing a classification model for smartphone data. The purpose of the classification model is to formally assess the authenticity of application-generated smartphone data residing on a smartphone. The output of the model is an authenticity classification – an ordered pair of values that expresses the grade of authenticity and the completeness of the evaluation.

The following sections describe the categorization of the requirements, the computation and representation of an authenticity score, the measurement of the completeness of an evaluation, along with the visualization of the final authenticity classification.

3.1   Categorization of the Requirements

Mathematical equations are required to consistently classify the authenticity of evaluated smartphone data. The equations must embody the requirements and smartphone data evaluation model presented in Section 2. In total, eleven requirements were identified and the evaluation of each requirement involves one or more assessment points. Each assessment point has one of three outcomes: (i) yes; (ii); no; or (iii) absent. A positive result of yes confirms that the requirement is met. A negative result of no indicates that the evaluated data does not meet the requirement. An absent result is assigned when the data is unavailable or insufficient.

The results produced by the assessment points are not equally important because each assessment point evaluates different aspects of the authenticity of smartphone data. The categorization of the assessment points into classes, each with a distinct focus, enables a more accurate evaluation of data authenticity.

Two classes are defined based on the notion of smartphone data authenticity considered in this work. Class A contains assessment points that confirm that no opportunity existed to change the smartphone data. Class B comprises assessment points that evaluate the consistency of the components responsible for creating smartphone data, as well as the consistency of the data itself. The assessment points in Class B evaluate the smartphone, smartphone applications and data associated with the applications. Therefore, Class B assessment points are placed in the following three subclasses:

- **Subclass B.1:** Assessment points that only evaluate application data.

- **Subclass B.2:** Assessment points that evaluate application behavior and the file structure used to store data.

- **Subclass B.3:** Assessment points that evaluate the smartphone state.

Figure 3 categorizes the assessment points according to the established classes and the core components involved in the requirements for authentic smartphone data. The categorization of the assessment points into Class A and Class B allows for weighted calculations of the authenticity scores.

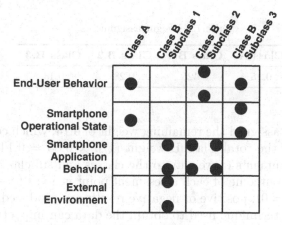

Figure 3. Categorization of assessment points.

3.2   Authenticity Score

The computation of the authenticity score is weighted because the outcome of each assessment point impacts the authenticity of the smartphone data differently. The weight assigned to each class should reflect the impacts that the constituent assessment points have on the final authenticity score.

Since Class A contains approximately 15% of the assessment points (Figure 3), a weight of 0.15 is assigned to the class. Class B, which contains the remaining assessment points, is assigned a weight of 0.85.

The Class B weight is subdivided to assign appropriate weights to its constituent subclasses. Subclass B.1 assessment points focus strictly on the evaluation of smartphone application data, which has a significant influence on the authenticity score. Since the Subclass B.1 assessment points are important, the subclass is assigned a weight of 0.425, one-half of the total weight of its parent Class B (0.85).

Assessment points in Subclass B.2 focus on the behavior of the smartphone application, but exclude the application data. Since these assessment points have less influence on the authenticity score than the Subclass B.1 assessment points that focus on data, Subclass B.2 is assigned a weight of 0.28, two-thirds of the remaining weight of Class B, which corresponds to one-third of the total weight of Class B ($1/3 \times 0.85 = 0.28$).

The assessment points in Subclass B.3 focus only on the smartphone and do not directly address smartphone applications and related data; thus, they have a limited influence on the authenticity score. Therefore,

Table 1. Weight assignments.

Class A	Class B.1	Class B.2	Class B.3
0.15	0.425	0.28	0.14

Subclass B.3 is assigned the remaining weight of 0.14, which corresponds to one-sixth of the total Class B weight ($1/6 \times 0.85 = 0.14$). Table 1 shows the assignments of weights to the classes and subclasses.

Because the outcome of each assessment point is a yes ($= +1$), no ($= -1$) or absent ($= 0$), positive or negative results are produced. However, the acquisition technique used to obtain the data can impact the ability to assess all the assessment points. Therefore, for each class c, the collection of positive results pos_c are divided by the number of assessment points n_c evaluated per class. The result is then weighted using the class weight w_c shown in Table 1.

Thus, the authentication score S_A for Class A is computed as:

$$S_A = w_c \frac{pos_c}{n_c} \tag{1}$$

The authentication score S_B for Class B is computed as the sum of the individual scores of its subclasses:

$$S_B = \sum_{c=B.1}^{B.3} w_c \frac{pos_c}{n_c} \tag{2}$$

The final authenticity score A_s is computed as the sum of the scores computed for Classes A and B:

$$A_s = \sum_{c=A}^{B} S_c \tag{3}$$

3.3   Authenticity Grading Scale

The authenticity score, as computed above, expresses the authenticity of the evaluated smartphone data as a percentage. The percentage value alone is inadequate – further description and categorization are required to better reflect the authenticity of smartphone data. Specifically, the categorization requires additional interpretation of the evaluated assessment points and all the possible outcomes. Since the number of assessment points evaluated and the possible outcomes factor significantly in the categorization of the authenticity score, it is necessary to confirm

Table 2.   Authenticity grading scale for smartphone data.

Grade	Description
Unsatisfactory	Fails to meet most of the requirements.
Low	Meets some of the requirements.
Moderate	Meets most of the requirements in Subclasses B.2 and B.3.
High	Meets most of the requirements in Subclasses B.1 and B.2.

the evaluations of the assessment points and compute all the possible outcomes relating to the evaluations of these assessment points. The result is a set of outcomes that has a normal distribution.

The normal distribution presents two clusters of potential outcomes. The first cluster (below the mean of the normal distribution) corresponds to the outcomes of the evaluated assessment points that mostly produce negative results. The outcomes are further grouped as follows:

- **Unsatisfactory Authenticity:** The outcomes of the evaluated assessment points produce only negative results.

- **Low Authenticity:** The outcomes of the evaluated assessment points produce negative results that outweigh the positive results.

The second cluster of outcomes (above the mean of the normal distribution) corresponds to the outcomes of the evaluated assessment points that mostly produce positive results. The outcomes are further grouped as follows:

- **Moderate Authenticity:** The outcomes of the evaluated assessment points produce positive results that outweigh the negative results.

- **High Authenticity:** The outcomes of the evaluated assessment points produce only positive results.

Table 2 shows the four grades in the authenticity grading scale. In order to assign a grade to the final authenticity score, it necessary to divide the normal distribution of all the outcomes into quartiles. The lower quartile distinguishes between the unsatisfactory and low authenticity grades, the middle quartile distinguishes between the low and moderate authenticity grades, and the upper quartile distinguishes between the moderate and high authenticity grades.

The quartiles enable the authenticity grading scale to provide context and better describe smartphone data authenticity. The quartiles create the boundaries between distinct grades of authenticity. The authenticity score is then plotted on the scale to determine the authenticity grade of the evaluated smartphone data. The consistent and formal measurement of smartphone data ensures that a digital forensic professional can conclusively establish the authenticity of smartphone data and also comprehend different grades of authenticity.

3.4   Completeness

The computation of authenticity scores and construction of the authenticity grading scale depend on the collection of assessment points that are evaluated. The specific acquisition technique used to obtain smartphone data strongly influences the availability of assessment points. A completeness score is required to express the number of the assessment points evaluated per component with respect to the number of available assessment points per component. This score would enable a digital forensic professional to present the completeness of the smartphone data evaluation with confidence, thereby complementing the authenticity grade.

The completeness score C_s is given by:

$$C_s = \sum_{i=1}^{4}(\frac{a_i}{t_i})(0.25) \tag{4}$$

where a_i is an evaluated assessment point and t_i is the total number of assessment points available for the component. For each component specified in Section 2.1, the evaluated assessment points a_i are counted and divided by the total assessment points t_i, yielding a weighted score computed using a 25% weight measurement per component. The weighted score ensures that each component is equally important. Evaluating a larger collection of assessment points would yield a more thorough classification of the authenticity of smartphone data. The availability of fewer assessment points would yield a partial evaluation, reducing the confidence in the authenticity of smartphone data.

3.5   Authenticity Classification

The authenticity A_S and completeness C_S scores are the key results produced by the classification model. The final authenticity classification A_C of the evaluated smartphone data is an ordered pair of the two individual scores:

$$A_C = \ <A_S; C_S> \tag{5}$$

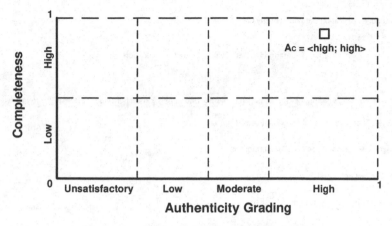

Figure 4. Authenticity classification graph.

The authenticity classification graph in Figure 4 shows a visual representation of the final authenticity classification. The x-axis represents the authenticity grading scale; the vertical lines divide the space into four quartiles corresponding to the four grades of authenticity. The y-axis represents the completeness scale; the single horizontal line distinguishes between high confidence and low confidence. The square in the top-right corner of the graph shows an example authentication classification of A_C = <high; high>.

4. Authenticity Classification Tool

A proof-of-concept tool was developed to automate the computation of the authentication classifications of smartphone data. Although a digital forensic professional could perform the computations manually, the automation eliminates human error and supports the visualizations of the results.

4.1 Tool Description

The tool computes and presents the authenticity classifications of evaluated smartphone data. Specifically, the tool supports the evaluation of all the assessment points of all the requirements. Note that each assessment point has one of three outcomes: yes (= +1), no (= −1) or absent (= 0). Equations 1 through 5 are used to compute an overall authenticity classification.

Figure 5 shows the user interface of the tool. The central viewing area has functional tabs, three interactive buttons and a canvas for rendering the authenticity classification graph. Each tab represents a com-

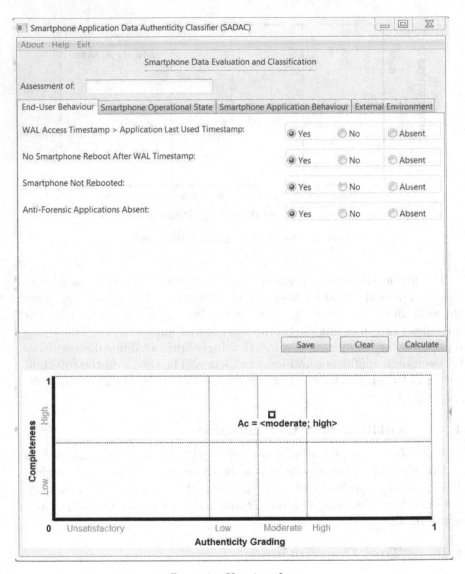

Figure 5. User interface.

ponent of authentic smartphone data and captures all the assessment points associated with the requirements for the component. Three radio buttons are provided to enter the outcomes for assessment points; the buttons ensure that only one option from yes, no and absent is selected for an assessment point. The Calculate button collects the results of all the evaluated assessment points and computes the authenticity classification. The final authenticity classification is presented within the authenticity classification graph in the canvas panel below the buttons.

4.2   Experimental Results

An experiment was conducted to validate the classification model. The experiment relied on a generic process for smartphone data manipulation [14]. The following four steps were involved in smartphone data manipulation:

- Ensure that the selected smartphone is accessible by confirming that the smartphone is either rooted (Android) or jailbroken (iOS).

- Select the application and identify the location of the files (e.g., SQLite database) that contain smartphone data.

- Identify the most appropriate approach for accessing smartphone data – either direct or off-device. The direct approach performs the manipulation of the smartphone data directly on the smartphone and relies on a program or utility to access the files. The off-device approach requires the files to be transferred to and from a connected computer with the required program or utility installed on the computer to perform the manipulation.

- Perform a manual reboot of the smartphone.

The experiment used an iPhone 7 as the test device. A new, albeit fabricated, text message was created on the device. A generic process for smartphone data manipulation was used to create the fabricated text message. The following steps were involved in creating the fabricated text message:

- Jailbreak the iPhone 7 using the extra_recipe + yaluX application.

- Pinpoint the storage structure (SQLite database) of the iPhone's default messaging application (/private/var/mobile/Library/ SMS/sms.db).

- Employ the direct approach and insert a fabricated text message in the SQLite database using the pre-installed sqlite3 command-line utility.

- Reboot the iPhone 7 to complete the manipulation process and ensure that the changes are reflected on the smartphone.

Table 3. Traces created by the experiment.

Trace	Trace Description
T_1	Automatic installation of the Cydia application
T_2	Unavailability of over-the-air updates
T_3	Discrepancies between write-ahead log file entries and application usage timestamps
T_4	Use of the `sqlite3` program
T_5	Presence of a clean write-ahead log file
T_6	Creation of entries in the reboot log file
T_7	Discrepancies in the mobile network provider records

The manipulation of the smartphone data has inherent side-effects that create various traces. Table 3 lists the traces specific to the experiment. Jailbreaking the iPhone 7 causes the automatic installation of the Cydia application and prevents over-the-air updates. Gaining access to the persistent data in the SQLite database via the direct approach, but without accessing the application, causes a discrepancy between the last modification timestamp of the SQLite database and the last usage timestamp of the application. The direct approach relies on the `sqlite3` program to gain access to the persistent data, which changes the last access timestamp associated with the program. This timestamp also closely follows the last modification timestamp of the SQLite database. Accessing the SQLite database to manipulate the record causes an immediate checkpoint to occur. Therefore, after closing the SQLite database, a clean and empty write-ahead log file is present on the iPhone 7. Finally, rebooting the iPhone 7 creates a new entry in the `/var/mobile/logs/lockdownd.log` reboot log.

Note that, although this was not observed in the case of the test iPhone 7, creating a fabricated text message causes discrepancies in the records captured by mobile network providers.

The traces listed in Table 3 were used to evaluate the authenticity of the smartphone data. The outcome of the authenticity grading is expected to be low or unsatisfactory due to the changes made to the iPhone 7 when implanting the fabricated text message. A high completeness value is anticipated because all the assessment points were evaluated.

Figure 6 presents the authenticity classification of the evaluated smartphone data. The computed authenticity classification confirms the assignment of a low authenticity grading. Furthermore, the authenticity classification also confirms a high completeness value, which is antici-

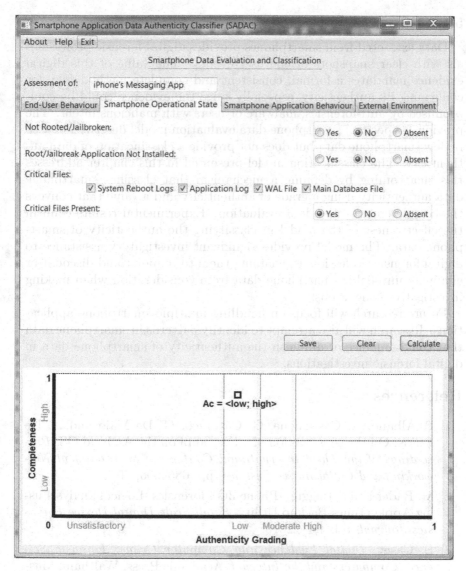

Figure 6.   Experimental results.

pated because all the assessment points of all requirements were evaluated. The assigned authenticity classification aligns with the predicted outcome and confirms that the manipulation does indeed influence the authenticity of the data.

5.   Conclusions

Data extracted from smartphones provides digital forensic professionals with clear snapshots of end-user events. The value of this digital evidence mandates a formal, consistent and complete methodology for confirming its authenticity, especially since the evidence could be compromised by anti-forensics, malware or users with malicious intent. The previously-specified smartphone data evaluation model describes how to review smartphone data but does not provide a classification of data authenticity. The classification model presented in this chapter addresses this shortcoming by defining a mechanism that classifies smartphone data authenticity using a grade of authenticity and a value that conveys the completeness of the data evaluation. Experimental results confirm the effectiveness of the model in classifying the authenticity of smartphone data. The model provides significant investigatory assistance to digital forensic professionals, enabling them to pinpoint and discount or eliminate unreliable smartphone data from consideration when making investigative conclusions.

Future research will focus on handling multiple smartphone applications. Research will also attempt to identify patterns in smartphone data that could enhance or diminish the authenticity of smartphone data in digital forensic investigations.

References

[1] P. Albano, A. Castiglione, G. Cattaneo, G. De Maio and A. De Santis, On the construction of a false alibi on the Android OS, *Proceedings of the Third International Conference on Intelligent Networking and Collaborative Systems*, pp. 685–690, 2011.

[2] M. Bader and I. Baggili, iPhone 3GS forensics: Logical analysis using Apple iTunes Backup Utility, *Small Scale Digital Device Forensics Journal*, vol. 4(1), 2010.

[3] E. Casey, *Digital Evidence and Computer Crime: Forensic Science, Computers and the Internet*, Academic Press, Waltham, Massachusetts, 2011.

[4] G. Cecere, N. Corrocher and R. Battaglia, Innovation and competition in the smartphone industry: Is there a dominant design? *Telecommunications Policy*, vol. 39(3-4), pp. 162–175, 2015.

[5] F. Cohen, *Digital Forensic Evidence Examination*, Fred Cohen and Associates, Livermore, California, 2009.

[6] L. Duranti, From digital diplomatics to digital records forensics, *Archivaria*, vol. 68, pp. 39–66, 2009.

[7] M. Hannon, An increasingly important requirement: Authentication of digital evidence, *Journal of the Missouri Bar*, vol. 70(6), pp. 314–323, 2014.

[8] R. Harris, Arriving at an anti-forensics consensus: Examining how to define and control the anti-forensics problem, *Digital Investigation*, vol. 3(S), pp. S44–S49, 2006.

[9] W. Jansen and R. Ayers, Guidelines on Cell Phone Forensics, NIST Special Publication 800-101, National Institute of Standards and Technology, Gaithersburg, Maryland, 2007.

[10] M. Losavio, Non-technical manipulation of digital data, in *Advances in Digital Forensics*, M. Pollitt and S. Shenoi (Eds.), Springer, Boston, Massachusetts, pp. 51–63, 2005.

[11] H. Pieterse, M. Olivier and R. van Heerden, Playing hide-and-seek: Detecting the manipulation of Android timestamps, *Proceedings of the Information Security for South Africa Conference*, 2015.

[12] H. Pieterse, M. Olivier and R. van Heerden, Evaluating the authenticity of smartphone evidence, in *Advances in Digital Forensics XIII*, G. Peterson and S. Shenoi (Eds.), Springer, Cham, Switzerland, pp. 41–61, 2017.

[13] H. Pieterse, M. Olivier and R. van Heerden, Smartphone data evaluation model: Identifying authentic smartphone data, *Digital Investigation*, vol. 24, pp. 11–24, 2018.

[14] H. Pieterse, M. Olivier and R. van Heerden, Detecting manipulated smartphone data on Android and iOS devices, in *Communications in Computer and Information Science*, H. Venter, M. Loock, M. Coetzee, M. Eloff and J. Eloff (Eds.), Springer, Cham, Switzerland, pp. 89–103, 2019.

[15] B. Schatz, Digital Evidence: Representation and Assurance, Ph.D. Thesis, Information Security Institute, Faculty of Information Technology, Queensland University of Technology, Brisbane, Australia, 2007.

Chapter 4

RETROFITTING MOBILE DEVICES FOR CAPTURING MEMORY-RESIDENT MALWARE BASED ON SYSTEM SIDE-EFFECTS

Zachary Grimmett, Jason Staggs and Sujeet Shenoi

Abstract   Sophisticated memory-resident malware that target mobile phone platforms can be extremely difficult to detect and capture. However, triggering volatile memory captures based on observable system side-effects exhibited by malware can harvest live memory that contains memory-resident malware. This chapter describes a novel approach for capturing memory-resident malware on an Android device for future analysis. The approach is demonstrated by making modifications to the Android `debuggerd` daemon to capture memory while a vulnerable process is being exploited on a Google Nexus 5 phone. The implementation employs an external hardware device to store a memory capture after successful exfiltration from the compromised mobile device.

Keywords: Mobile device malware, system side-effects, memory capture

1.   Introduction

Mobile devices are increasingly being used to process and manage personal and sensitive information such as photos, videos, browsing history, notes, social media posts and bank account data. As a result, these devices have become attractive targets for adversaries and attacks on the devices are increasing in their scope and magnitude [5].

Mobile devices share several attack vectors with traditional workstations (e.g., Wi-Fi and Bluetooth adapters). However, mobile devices are also continuously connected to cellular networks in which the device owners have little to no control. Fragmentation in mobile device operating systems and embedded device architectures makes it difficult to

© IFIP International Federation for Information Processing 2019
Published by Springer Nature Switzerland AG 2019
G. Peterson and S. Shenoi (Eds.): Advances in Digital Forensics XV, IFIP AICT 569, pp. 59–72, 2019.
https://doi.org/10.1007/978-3-030-28752-8_4

develop exploits that impact multiple devices, but it also renders the mobile device ecosystem more challenging to secure.

Vulnerabilities that affect large families of devices have been demonstrated [1, 3, 4, 12]. These vulnerabilities make it imperative that new efforts be developed to secure mobile devices against increasingly sophisticated attacks. Analyzing and understanding the rapidly evolving threats to mobile devices require the capture and analysis of evidence pertaining to attacks on the devices.

Memory-resident malware is difficult to detect because it resides entirely in volatile memory and does not write to secondary memory. Additionally, this type of malware often removes itself from memory after execution. This makes it impossible for a forensic analyst to identify and collect malware after a compromise has occurred. The only option is to take proactive measures to capture the contents of memory while the malware still resides in memory. In addition to supporting forensic investigations, the ability to capture the malware enables researchers to identify and mitigate the vulnerabilities exploited by the malware.

Mobile devices are exposed to unique threats compared with stationary devices (e.g., workstations) because of their mobility. Moreover, real-world mobile devices incorporate peripherals such as communications processors that are not present in most virtual or emulated devices. Therefore, it is important to leverage real-world mobile devices to understand and mitigate the unique threats.

The proposed approach leverages digital forensic and embedded device engineering techniques to capture evidence of malicious activity on mobile devices [8]. Consumer hardware, specifically a Google Nexus 5 smartphone, was adapted to capture transient malware, and multiple techniques for storing the captured information are evaluated. The Stagefright family of exploits is used as a case study to explore and identify strategies for detecting various types of malware.

2.   Malware Categorization

Security monitoring solutions typically rely on identifying malware based on artifacts (e.g., data or code) that reside in a filesystem and/or by examining how malware behaves during execution [2, 7]. Malware is identified by developing and checking for storage signatures corresponding to malware artifacts and/or execution signatures that describe malware behavior. Malware developers attempt to elude signature-based detection by making slight modifications to malware code and/or behavior. In turn, malware analysts attempt to generalize the storage and execution signatures to detect variations of the same malware. Although these approaches may work to varying degrees for known malware, they cannot be applied effectively to (unknown) malware that has not been studied previously.

Grimmett et al. [6] have proposed alternative methods for identifying malware based on observable system side-effects. They also present a taxonomy for categorizing malware according to its behavior and system side-effects. The taxonomy covers three categories of malware based on: (i) user-detectable behavior; (ii) system-detectable behavior; and (iii) inconspicuous behavior. Each malware category exhibits different characteristics that can be leveraged to develop system side-effect signatures for detecting and capturing the malware in question.

Grimmett et al. [6] also present a case study involving the Stagefright malware. Stagefright is designated as system-detectable malware because it produces side-effects that are detectable by the underlying operating system (i.e., Android). The system side-effects are a result of repeated attempts at exploiting a system service that causes a service to crash (i.e., brute force execution). Due to the reliability requirement imposed on mobile devices, critical services automatically restart after a crash (e.g., due to a failed exploit attempt), enabling an attacker to attempt to exploit the vulnerability in the system service again. In some instances, the crashed system service is transparent to the end-user; this enables an attacker to attempt the exploit repeatedly until it succeeds and without alerting the user.

The crashing of a service as a result of a failed exploit attempt is, in fact, a side-effect that is observable to the underlying operating system. As a result, this side-effect can be used to trigger events that could assist a malware analyst in identifying and collecting previously unknown malware.

2.1 Stagefright

The Stagefright family of vulnerabilities include integer overflows and heap overflows deep in the MPEG4 media processing portions of the libstagefright Android operating system library [12]. The vulnerabilities are critical because they can be triggered remotely by sending specially-crafted MMS messages to mobile device users. The entire exploitation process is transparent to a user in that it does not trigger warnings or error messages.

In order for a Stagefright exploit to succeed, it has to defeat address space layout randomization (ASLR). Address space layout randomization is a memory protection mechanism supported by most modern operating systems to mitigate memory corruption exploitation attempts.

Address space layout randomization attempts to randomize the base addresses of key components of a process (e.g., libraries, the stack and the heap) to make it more difficult for an attacker to reliably jump to a known piece of code in memory.

To overcome this barrier, Stagefright guesses the locations of the base addresses of the libstagefright library in the mediaserver process. When the locations are guessed incorrectly, the Android mediaserver process simply crashes and restarts, reloading the process into the same vulnerable state. Thus, multiple exploitation attempts and subsequent crashes tend to occur when a Stagefright exploit is executed. Since the mediaserver process runs at a privileged level, successful exploitation of the libstagefright library in the mediaserver process enables the attacker to inherit system-level permissions. These characteristics make Stagefright an excellent candidate for demonstrating that system side-effects produced by malware can be used to capture the malware while it is still in volatile memory. As a result, the Stagefright family of vulnerabilities is considered as a case study in this research.

2.2   Live Memory Analysis

A number of tools have been developed for acquiring memory images from volatile memory (i.e., RAM) [9–11]. A widely used open-source tool is Linux Memory Extractor (LiME). LiME is a loadable Linux kernel module that can dump the entire physical memory of a device. In an attempt to be forensically sound, LiME is designed to have a very limited memory footprint. These characteristics make LiME a useful tool for collecting evidence of malicious activity that cannot be precisely located in memory.

Certain challenges must be addressed in order to use LiME to capture malware. First, LiME has minimal impact on the target system. Since LiME does not halt the system, it is necessary to ensure that the downloaded malware remains in memory when the capture process executes. Second, the memory image produced by LiME is the size of the device physical memory – this is about 2 GB in the case of a Google Nexus 5 device. The memory image file can be stored on device (local) storage or saved over a TCP connection to a remote machine. When network access is not available, the number of captures that can be stored are limited by the amount of storage space available on a mobile device.

While LiME is ideal for collecting large amounts of memory at a given time, it is not the best choice for consistent or continuous monitoring of live memory. This makes LiME useful in situations where it can be invoked when suspicious activity such as a system-detectable side-effect is

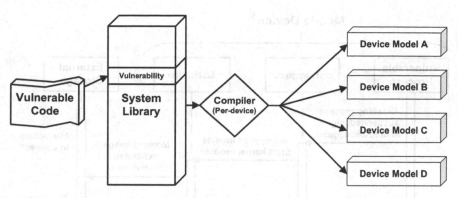

Figure 1. Proliferation of a vulnerability in a system library.

detected. The case study described in this chapter focuses on capturing malware from mobile devices. However, the captured malware is only useful if analysts can examine and understand what has been captured.

3. Automated Memory Acquisition

The mobile device malware analysis community lacks a mechanism for reliably and feasibly capturing a snapshot of memory during system exploitation attempts. This section describes a proof-of-concept implementation that demonstrates the viability of automated memory acquisition from an Android mobile device. The proof-of-concept has been implemented on a Google Nexus 5 phone. By modifying the Android debuggerd daemon, the physical memory contents are dumped upon invoking the LiME kernel module during the crash of a system process. Because of the limited storage on the mobile device, the memory capture is subsequently exfiltrated to another device using TCP via USB forwarding.

3.1 Design Requirements

This research was motivated by the concern that a vulnerability in a system library (e.g., Stagefright) puts a large number of mobile devices at risk for remote exploitation [4, 12]. The focus on system library vulnerabilities is important. This is because, to maximize the impact, malware developers invest resources in identifying vulnerabilities and exploits in the system libraries of popular operating systems.

Figure 1 demonstrates how a vulnerability in a system library becomes a vulnerability for every device that uses the library. Additionally, since services traditionally execute with higher privileges than user applications, attackers have an additional incentive to exploit the services.

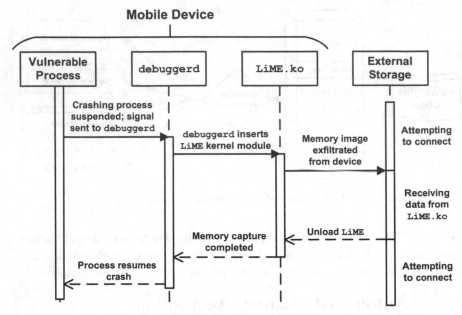

Figure 2.   Malware capture process.

The software modifications that support live memory acquisition from a mobile device should be reliable and should have minimal impact on observable device behavior. The modifications must also run on a physical device so that all the potential vectors are available for study. The hardware should require as few proprietary modifications to the operating system as possible. Additionally, the modifications should be adaptable to newer devices and future operating system versions in order to meet future malware analysis needs.

3.2   Implementation and Testing

Figure 2 presents the process for acquiring a memory image from a mobile device after a service has crashed. When a service crashes, the debuggerd handler is signaled, which suspends the crashing service. The debuggerd daemon also initializes the LiME kernel module and specifies the capture parameters (e.g., image format and exfiltration method). The primary use case for this implementation is to transfer the image using TCP via USB forwarding; however, transfer via Wi-Fi is also supported by the implementation. Additionally, the acquired memory image may be moved to device local storage for transfer at a later time.

The unmodified debuggerd daemon suspends all crashing processes when it generates a tombstone file – it is after debuggerd has completed its own crash handling functionality that the process is left suspended and gdb is attached or the process is allowed to continue and crash. Because debuggerd suspends a process while gathering information, additional code added to debuggerd can execute before the process is allowed to resume. After the image acquisition is complete, the kernel module is unloaded and debuggerd allows the process to resume and crash. The malware capture hardware stores the acquired image and proceeds to wait for another memory acquisition; the crashed process may then be restarted by the operating system.

3.3   Android Modification Results

A framework was created to manage a (mobile) device-under-test and enable automated testing. The framework, which was developed in Python, creates test instances that use the Android Debug Bridge (adb) to interface with Android devices. These test objects can be extended to create new test objects with additional functionality as desired. The tests were extended to enable automated testing of the LiME kernel module and to verify that the acquired memory samples contained the target crash vectors.

The crash vector was recovered from memory in order to determine if the entire vector was captured successfully. The Volatility plugin linux_pslist was used to determine the mediaserver process identifier. Next, linux_yarascan was used to search the virtual memory of the process for ftyp, which denotes the "File Type Box" that appears in the beginning of some MPEG4 media files (e.g., crash vector). Next, linux_proc_maps was used to determine the mapped memory sections that needed to be extracted for analysis, upon which linux_dump_map created an image of the relevant memory mapping from the mediaserver process. A Python script was written to verify that the dumped memory contained the crash vector.

The LiME kernel module is designed to provide minimally-invasive memory acquisition for forensic analysts. The proof-of-concept implementation does not assume that the device is handled using digital forensic best practices. Therefore, additional testing had to be conducted to verify that data remains in memory long enough to be captured using LiME. Additionally, any differences between suspended and non-suspended processes had to be understood to determine the impact of suspension on memory acquisition.

To determine if LiME was suitable for the proposed tasks, tests were conducted to measure how effectively LiME captures an MPEG4 crash vector when it is executed manually immediately following a browser

Table 1.   Crash vector capture success rates during manual testing.

	No Wait after Reboot		Wait after Reboot	
	Local Storage	TCP Capture	Local Storage	TCP Capture
Suspended	25%	0%	100%	100%
Not Suspended	0%	50%	0%	100%

crash. The tests were performed using a capture to local device storage and exfiltration via TCP over a USB connection to an external host. Additionally, tests were performed with and without processes set to suspend and wait after a crash. Moreover, the tests were executed with and without a one-minute wait between restarting the device and performing the crash and memory acquisition.

The results in Table 1 demonstrate the impact of a short wait on the success rate. The wait/no-wait results demonstrate that the target data is likely to be lost unless it is captured quickly or the process is suspended. In the experiment, the device-under-test was restarted before every memory acquisition test to ensure that no residual data remained from previous tests. During the Android startup process, the user interface was made available as quickly as possible and other startup tasks were executed in the background. Because the background startup operations were still initializing the system, memory was released and reused more rapidly than under normal operating conditions. Therefore, the startup period had to be allowed to complete or the memory acquisition would likely be disrupted by the high memory turnover.

Figure 3 shows the methodology for testing the reliability of capture of an MPEG4 crash vector using the proof-of-concept implementation. The device-under-test navigates to the crash vector (`crash.mp4`) on a local webserver, which causes `mediaserver` to crash. When `mediaserver` crashes, `debuggerd` handles the crash and inserts the `LiME` kernel module and performs the memory acquisition. The acquired memory image is then searched for instances of the known crash vector.

A successful capture includes at least one complete and intact instance of the crash vector. If the complete vector cannot be found in memory, it may be possible to find partial instances. The partial instances would be less valuable than a complete sample from the perspective of malware analysis. However, further investigation may enable a complete instance to be reconstructed from memory.

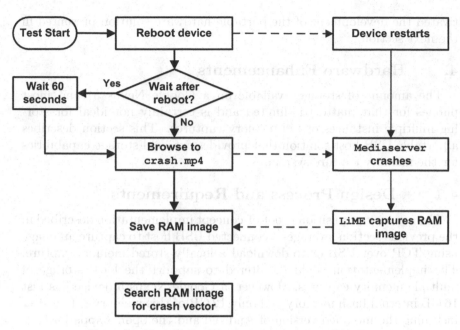

Figure 3. Test design for capture technique validation.

When LiME acquires memory and exfiltrates the image to a remote host via TCP, the rate of capture is limited by the network bandwidth between the device and remote host. The proof-of-concept implementation was designed to support capture via a Wi-Fi network. However, as discussed below, the time required to complete the capture limits the effectiveness of this approach.

Table 2. Average capture time for the exfiltration methods.

Exfiltration Method	Time (seconds)
Local Storage (Capture Only)	59.79
Local Storage and USB Downloading	446.95
TCP Exfiltration via USB Forwarding	382.22
TCP Exfiltration via Wi-Fi Network	2,382.93

Table 2 shows the average time required for various memory acquisition techniques. Note that the testing framework incorporates approximately 5% additional overhead for network operations using Python sockets, which is not enough to disrupt the experimental results. The significant increase in time required to acquire an image over Wi-Fi motivated the development of the portable hardware solution presented in the next section.

4.   Hardware Enhancements

The amount of storage available on a Google Nexus 5 (and most phones for that matter) is limited and is certainly not ideal for storing multiple instances of full memory captures. This section describes a portable USB host solution that provides external storage capabilities for the memory capture system.

4.1   Design Process and Requirements

The memory acquisition proof-of-concept implementation described in the previous section leverages a connected USB host to capture an image using TCP over USB or to download a locally-stored memory capture. The implementation could be altered to support the local storage of multiple memory captures. However, a Google Nexus 5 device has just 16 GB internal flash memory and only 12 GB of this memory is free after installing the modified version of Android and the Open GApps package that contains Google Chrome. Additionally, a Google Nexus 5 does not support any removable storage (e.g., microSD card). Because each captured memory image is 2 GB in size, the number of captures that can be stored at one time is severely limited.

An external storage solution also reduces the likelihood of a captured memory image being erased or corrupted. Because this research has focused on malware capture for future analysis, it would not be prudent to rely on a compromised mobile device to preserve the captured image. Furthermore, since a USB connection to a compromised device could put the USB host at risk, the external storage solution should be easily wiped and redeployed as necessary.

It is also important that the hardware support package be portable. This requires the external storage solution to incorporate a battery, which imposes a limit on the length of time the hardware package can be used between charges. Thus, the portable hardware should consume as little power as possible while remaining reliable and available.

4.2   Implementation Details

Figure 4 shows the proof-of-concept implementation created using a Raspberry Pi 3 as a USB host. The Raspberry Pi 3 runs the Raspbian Jessie Lite operating system (a minimal operating system based on Debian Jessie) and includes adb binaries compiled with the same toolchain used by the software modification proof-of-concept system de-

Figure 4. Memory capture support hardware.

scribed above. The Raspberry Pi has an 802.11n wireless radio and multiple USB ports that enable it to support memory captures over Wi-Fi or USB.

The Raspberry Pi 3 is a general-purpose computing device that uses more power than a dedicated microcontroller. However, the availability of a Linux operating system enables the memory capture device to be more adaptable than an embedded device. The increased power requirement is a reasonable trade-off for the additional functionality and ease-of-use provided by the operating system. Specifically, the operating system enables the memory capture device to incorporate logic that controls the behavior of the capture software and determine when to download a completed local capture. Additionally, the device hosts an SSH server that enables the device to be remotely operated and configured.

The only way to safely shut down the Raspberry Pi is via the `shutdown` or `halt` commands – disconnecting the device from power without shutting it down properly could corrupt the microSD card and the captured memory images it contains. The implementation incorporates an Anker Power Bank with 8,400 mAh capacity and an external charge indicator. The external charge indicator should be monitored to minimize the risk of draining the battery and corrupting the captured memory images.

An alternative solution is to use a second mobile device to support the memory capture device. Using an Android device would eliminate the need for an external battery while enabling similar capabilities as a Raspberry Pi (i.e., Linux operating system and `adb` support). However,

Table 3. Average download times of memory images via adb.

ADB Host	Time (seconds)
MacBook Pro (2 GHz Intel Core i7)	317.40
Raspberry Pi 3 (1.2 GHz Cortex-A53)	422.50

the choice of mobile device is limited by the same constraints that motivate the use of a support device – that is, the device would need to provide substantial external storage. In any case, a Raspberry Pi costs less than any similar mobile device.

4.3   Experimental Results

The Raspberry Pi 3 has a less powerful processor than the workstations used to test the proof-of-concept memory capture implementation. As a result, the portable storage solution requires more time to perform the tasks than the times listed in Table 2. Table 3 presents the times required to download locally-stored memory images via adb.

Device power usage was measured using a USB power monitor between the battery and Raspberry Pi. The power monitor measured the total power consumed by the device and provided instantaneous current and power measurements.

Table 4. Power consumption of the support hardware.

Device Status	Power (Amps)
Device idle; no connection	0.26
Device idle; phone connected with screen off	0.35
Device idle; phone connected with screen on	0.69
Device downloading; phone connected with screen on	0.70

Table 4 lists the instantaneous current measurements recorded during various states of device operation. When the mobile phone was connected to the Raspberry Pi, it began charging and drew additional power from the battery. This unintended side-effect caused the battery to drain faster than expected. However, the battery provided several hours of operation after it was fully charged.

The Raspberry Pi 3 schematics are limited and do not include the USB controller and connections. Previous versions of the Raspberry Pi have direct connections between the power input 5 V line and the 5 V line on the USB ports. However, the 5 V lines on the USB ports of the Raspberry Pi 3 are not powered when the device is powered without a bootable image available. This suggests that the USB controller may be able to disable the power output on the USB ports. Disabling the unnecessary power drain through the Raspberry Pi is not critical, but it would be useful for future applications of the hardware solution.

5. Conclusions

Mobile devices have complex attack surfaces and vulnerabilities that can be exposed and exploited when they connect to networks. Increasing device complexity and ubiquitous mobile access necessitate the development of new techniques for detecting and mitigating mobile device malware.

Sophisticated malware uses a variety of techniques to avoid detection and capture. Encryption and encoding have been used to evade signature-based detection for years. Self-destructing malware erases itself to avoid discovery during digital forensic investigations. Memory-resident malware that never uses non-volatile storage disappears when the device is shut down or rebooted.

These sophisticated malware features require novel detection and capture techniques. This chapter has described a new technique that enables the capture of memory-resident malware using live memory digital forensic tools (e.g., LiME). The automated capture technique enables the discovery and analysis of previously unknown exploitation techniques as well as the implementation of new mitigation strategies for vulnerable devices. Most importantly, the modifications required to implement the technique are minimal – the modified device contains the same vulnerabilities found in an unmodified version of the device.

A memory capture technique will not mitigate any vulnerabilities unless the captured malware can be analyzed successfully. Therefore, the capture technique is designed to support malware analysis. The captured images are compatible with the Volatility framework.

Future research will focus on developing improved guidelines and techniques for identifying malware in captured memory images. Additionally, memory images from normal devices and exploited devices will be compared in an attempt to automate malware analysis.

References

[1] H. Be'er, Metaphor: A (Real) Real-Life Stagefright Exploit, Revision 1.1, NorthBit, Herzliya, Israel (raw.githubusercontent.com/NorthBit/Public/master/NorthBit-Metaphor.pdf), 2016.

[2] R. Bejtlich, *The Tao of Network Security Monitoring: Beyond Intrusion Detection*, Addison-Wesley, Boston, Massachusetts, 2004.

[3] M. Brand, Stagefrightened? Project Zero, Google, Mountain View, California (googleprojectzero.blogspot.com/2015/09/stagefrightened.html), September 16, 2015.

[4] J. Drake, Stagefright: Scary code in the heart of Android, presented at the *Black Hat USA Conference*, 2015.

[5] G Data Software, G Data Mobile Malware Report, Threat Report: Q2/2015, Bochum, Germany, 2015.

[6] Z. Grimmett, J. Staggs and S. Shenoi, Categorizing mobile device malware based on system side-effects, in *Advances in Digital Forensics XIII*, G. Peterson and S. Shenoi (Eds.), Springer, Cham, Switzerland, pp. 203–219, 2017.

[7] C. Pfleeger and S. Lawrence-Pfleeger, *Security in Computing*, Prentice Hall, Upper Saddle River, New Jersey, 2007.

[8] Scientific Working Group on Digital Evidence, SWGDE Best Practices for Mobile Phone Forensics, Version 2.0, 2013.

[9] H. Sun, K. Sun, Y. Wang, J. Jing and S. Jajodia, TrustDump: Reliable memory acquisition from smartphones, *Proceedings of the Nineteenth European Symposium on Research in Computer Security*, part I, pp. 202–218, 2014.

[10] J. Sylve, A. Case, L. Marziale and G. Richard, Acquisition and analysis of volatile memory from Android devices, *Digital Investigation*, vol. 8(3-4), pp. 175–184, 2012.

[11] V. Thing, K. Ng and E. Chang, Live memory forensics of mobile phones, *Digital Investigation*, vol. 7(S), pp. S74–S82, 2010.

[12] Zimperium zLabs, The Latest on Stagefright: CVE-2015-1538 Exploit is Now Available for Testing Purposes, San Francisco, California (blog.zimperium.com/the-latest-on-stagefright-cve-2015-1538-exploit-is-now-available-for-testing-purposes), September 9, 2015.

Chapter 5

A TARGETED DATA EXTRACTION SYSTEM FOR MOBILE DEVICES

Sudhir Aggarwal, Gokila Dorai, Umit Karabiyik, Tathagata Mukherjee, Nicholas Guerra, Manuel Hernandez, James Parsons, Khushboo Rathi, Hongmei Chi, Temilola Aderibgbe and Rodney Wilson

Abstract   Smartphones contain large amounts of data that are of significant interest in forensic investigations. In many situations, a smartphone owner may be willing to provide a forensic investigator with access to data under a documented consent agreement. However, for privacy or personal reasons, not all the smartphone data may be extracted for analysis. Courts have also opined that only data relevant to the investigation at hand may be extracted.

This chapter describes the design and implementation of a targeted data extraction system for mobile devices. It assumes user consent and implements state-of-the-art filtering using machine learning techniques. The system can be used to identify and extract selected data from smartphones in real time at crime scenes. Experiments conducted with iOS and Android devices demonstrate the utility of the targeted data extraction system.

Keywords: Mobile devices, privacy, targeted data extraction, iOS, Android

1.   Introduction

Smartphones contain large amounts of data that are of significant interest in forensic investigations. However, these devices have in essence become personal data repositories and the privacy of their data is a serious concern. A landmark 2014 ruling by the U.S. Supreme Court in Riley v. California and subsequent rulings based on this case suggest that it may not be enough to obtain a warrant to conduct a search of a smartphone, but it may also be required to restrict the search to specific items on the device that relate to the crime being investigated. What is needed

© IFIP International Federation for Information Processing 2019
Published by Springer Nature Switzerland AG 2019
G. Peterson and S. Shenoi (Eds.): Advances in Digital Forensics XV, IFIP AICT 569, pp. 73–100, 2019.
https://doi.org/10.1007/978-3-030-28752-8_5

is a forensically-sound system that can perform targeted (selective) data extraction under a documented consent agreement. Commercial tools such as Cellebrite UFED Physical Analyzer have great utility, but they do not support targeted data extraction.

This chapter describes the design and implementation of a prototype software system that supports targeted data extraction from iOS and Android devices in a forensically-sound manner. The system runs on a solid state drive connected to a laptop, which is connected to a mobile device of interest on which the targeted data extraction app is downloaded. Metadata and content filtering rules in the app support targeted data extraction under a consent agreement signed by the device owner. Metadata filtering rules enable data of specific types with relevant creation dates/times and locations to be extracted. Content-based filtering leverages machine learning to exclude non-relevant data and ensure that user data privacy is maintained. Forensic soundness is realized using the eDiscovery Reference Model [19] and dynamic/live analysis techniques drawn from network and cloud forensics [17, 26].

2. Related Work

Several tools support full data acquisition from iOS and Android devices. Commercial tools include Cellebrite UFED Physical Analyzer, Paraben Electronic Evidence Examiner, Oxygen Forensic, AccessData Mobile Phone Examiner Plus, Microsystems XRY, Magnet Acquire and Blackbag Mobilyze. These tools attempt to acquire as much data as possible via logical and physical acquisitions. However, they do not support on-device or off-device selective methods for extracting only the data that is relevant to investigations.

Considerable research has focused on forensic data extraction and analysis. Some of this work deals with the extraction of specific types of artifacts from cloud drives and social networking applications [2, 4, 26]. Other research has been directed at general forensic data extraction techniques for mobile devices [14, 25]. Interested readers are referred to [23] and [28] for detailed discussions about iOS and Android device forensics, respectively.

The concept of "real-time triage" has become increasing important and there has been some work on building such systems [9, 27]. Another important aspect is data privacy in the context of digital forensics in general and mobile forensics in particular [3, 31].

Machine learning (see, e.g., [24]) and its applications have gained considerable attention in recent years. Deep learning (see, e.g., [20]) has been successfully applied in areas ranging from image recognition [18] to natural language translation [10]. Open-source frameworks such as Caffe [15], Theano [8] and TensorFlow [1] have been developed for implementing deep representational learning using neural networks. State-of-the-art processors in modern smartphones make it feasible to perform image analysis and classification, including facial detection, using deep learning models such as Inception [33], Open NSFW [22] and MobileNet [13].

At this time, mobile device forensic tools are unable to perform on-device targeted data extraction as described in this chapter. In fact, the available tools only extract images of device content and enable the images to be queried and analyzed in an off-device manner. Moreover, these tools do not have the ability to filter data using machine learning techniques.

3. System Overview

The targeted data extraction system (TDES) for mobile devices has three components: (i) data identification system; (ii) data acquisition system; and (iii) data validation system.

- **Data Identification System:** The data identification system is responsible for identifying the relevant files based on metadata and content. Input to the system is broadly driven by a consent form and is fine-tuned by the forensic investigator using a specially-designed user interface.

  Smartphone data comes in a variety of types. The basic categories of smartphone data are photos (images), videos, messages and contact lists. Each category is associated with metadata that describes aspects of the data, such as time (when an image was placed on the device), location (where the image was taken) and sender and receiver (of text and multimedia messages).

  Note that metadata is different from content. For example, a query based on a date range – "photos taken within the past week" – uses metadata about photos. However, a query for photos containing "weapons" would require content-based filtering. The data identification system incorporates state-of-the-art machine learning, natural language processing and data mining algorithms to perform content-based filtering.

- **Data Acquisition System:** The data acquisition system interacts with the data identification system to retrieve targeted files from a smartphone in a forensically-sound manner. Data acquisition corresponds to data collection; therefore, the data that is acquired is the desired evidence.

The data acquisition system incorporates two components: (i) TDES manager; and (ii) TDES app. The TDES manager is a system-on-chip that resides on a portable bootable drive. The manager boots up in Windows 10 when connected to a laptop or workstation. The target smartphone is connected to the same laptop or workstation in order to deploy the TDES app on the target smartphone. The user interface of the TDES app enables an investigator to provide input to the data identification system. Finally, the filtered data from the target phone is transferred to the TDES manager.

- **Data Validation System:** The data validation system, which is integrated with the data identification and data acquisition systems, ensures that data is transferred in a forensically-sound manner. It performs appropriate hashing to insure data integrity. Additionally, it generates a log timeline that documents all the steps taken by the TDES system during "live analysis." Finally, the data validation system produces a report that documents the needs of the investigator (e.g., queries), the data analysis that was performed and the data that was selected.

The data identification and data acquisition systems are described together because their abstractions are closely coupled. Also, because the system only performs logical data extractions, it is assumed that relevant data is not stored in hidden or deleted files. Furthermore, the focus is on rapid targeted data extraction – how to define what data is to be extracted, how to ensure that data extraction is done in a forensically-sound manner and how to perform data extraction very rapidly.

4.   Targeted Data Extraction

In order to motivate the development of the model for targeted data extraction, it is instructive to present potential application scenarios. These scenarios, which were suggested by forensic investigators, involve instances where consent is natural and the ability to filter data would be very useful:

- A car accident where a bystander has taken photos or a video of the incident.

- A drug overdose incident where the victim's phone has information about drugs and drug dealers.

- A suicide case where the victim's phone may contain relevant texts, email and photos.

- A domestic violence situation where the victim's phone has photos that document the physical abuse.

- A major incident where several individuals have captured videos and photos of the perpetrators, their weapons and their vehicles.

In several shooting incidents, bystanders and/or companions have recorded the events on their phones [30]. The Boston Marathon bombing case had a massive amount of digital evidence from multiple sources [32]. In these and many other incidents, automated selective data extraction would have been very useful.

Data of value in forensic investigations is classified as follows:

- **User-Created Data:** This includes contacts and address books, SMS messages, MMS messages, calendars, voice memos, notes, photographs, video/audio files, maps and location information, voice mail and stored files.

- **Internet-Related Data:** This includes browsing histories, email and social networking data.

- **Third-Party Application Data:** This includes messaging data (text, voice, video and pictures) from applications such as Facebook, WhatsApp and Skype.

As discussed above, the TDES app, which is deployed on the target device, is responsible for filtering and transferring the data to the TDES manager. This method of data extraction is called "on-device acquisition." In this type of acquisition, only the data that is filtered by the TDES app is transferred from the phone. No other data on the device is ever pushed to the TDES manager.

However, for some iPhone data types, it is not possible to selectively extract relevant data without "jailbreaking" the phone or using the iTunes backup system. Since jailbreaking is not employed in this work, the only option is to use the iTunes backup system. Selective data extraction from iTunes is referred to as "backup acquisition." In backup acquisition, all the available data from the iTunes backup is moved to the TDES manager, which extracts the relevant data and deletes the backup after the extraction is completed.

Table 1.   On-device metadata-based extraction.

Data Category	Metadata Type	iOS	Android
Photos	Date and time	Yes	Yes
Photos	Location	Yes	Yes
Photos	Album type	Yes	Yes
Videos	Date and time	Yes	Yes
Videos	Location	Yes	Yes
Contacts	Name	Yes	Yes
Contacts	Number	Yes	Yes
Contacts	Area code	Yes	Yes
Contacts	Email	Yes	Yes
Calendar Events	Date	Yes	Yes
Reminders	Date	Yes	Yes
Photos	Third-party app	No	Yes
Messages/SMS/MMS	Date and time	No	Yes
Messages/SMS/MMS	Contact number	No	Yes
Call Logs	Incoming call	No	Yes
Call Logs	Outgoing call	No	Yes
Call Logs	Missed call	No	Yes
Call Logs	Date and time	No	Yes
Notes	Search string	No	No
Notes	Date and time	No	No
Voice Memos	Date and time	No	No
Web History	Date and time	No	No
Email	Date and time	No	No
Facebook Messages	Date and time	No	No
WhatsApp Messages	Date and time	No	No
LinkedIn Messages	Date and time	No	No
WeChat Messages	Date and time	No	No
Viber Messages	Date and time	No	No

4.1   On-Device Metadata-Based Filtering

Table 1 shows the data that can and cannot be extracted by the TDES app in the on-device mode via metadata filtering. The first part of the table shows the data that can be extracted from iPhones (iOS devices) and Android phones. The second part of the table shows the data that can be extracted from Android phones, but not from iPhones (e.g., photos captured by third-party apps such as Facebook and WhatsApp). The third part of the table shows data that the TDES app currently cannot extract from iPhones and Android phones.

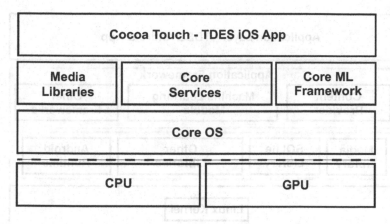

Figure 1. iOS frameworks.

- **iPhones:** System interfaces for iPhones are delivered in the form of packages called frameworks (Figure 1). The TDES app for iPhones uses several frameworks in Media Libraries and Core Services. The Photos framework provides direct access to photo and video assets managed by the iPhone Photos app. The AVKit framework provides a high-level interface for playing video content. The CoreLocation framework provides location and orientation information. The EventKit framework provides an interface for accessing calendar events. The Contacts framework provides access to user contacts and functionality for organizing contact information.

- **Android Phones:** Figure 2 shows the Android operating system stack. The TDES Android app, which is deployed in the application layer, leverages services provided by the Application framework, which includes the Content Provider, Activity Manager, Resource Manager and View [12]. Content Provider provides access to a range of data and other services used for design and implementation.

4.2   On-Device Content-Based Filtering

Trained machine learning models are developed using supervised learning techniques, including learning using deep neural nets. A trained model can be incorporated in the iOS or Android TDES app using the appropriate framework. The model can be used

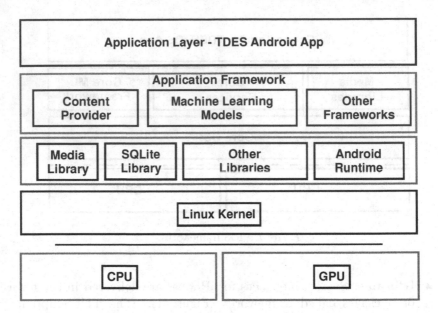

Figure 2.   Android operating system stack.

directly by retraining the final layer or by using heuristics based on model outputs.

The current versions of the TDES app employ adapted trained models from Inception-v3 [16], MobileNet [13] and Open NSFW [22] to classify photos and videos. Interested readers are referred to the bibliography for details about the accuracy of these models. The TDES apps are able to identify photos containing weapons, people, vehicles, drugs, websites, skin exposure and gadgets. The accuracy of the adapted models is discussed in Section 5.

The Core ML framework [5] is used for on-device content-based filtering on iPhones. Core ML provides support for several machine learning frameworks, including Vision and GameplayKit.

The TensorFlow Lite framework [35] is used for on-device content-based filtering on Android phones. The trained model and related labels are used in conjunction with a shared object file libtensorflow_inference.so, which is written in C++. The Java API libandroid_tensorflow_inference_java.jar [1, 29] is used to interface with Android platforms.

Table 2. Off-device metadata-based extraction.

Data Category	Metadata Type	iOS	Android
Photos	Date and time	Yes	Yes
Photos	Location	Yes	Yes
Photos	Album type	Yes	Yes
Videos	Date and time	Yes	Yes
Videos	Location	Yes	Yes
Contacts	Name	Yes	Yes
Contacts	Number	Yes	Yes
Contacts	Area code	Yes	Yes
Contacts	Email	Yes	Yes
Calendar Events	Date	Yes	Yes
Reminders	Date	Yes	Yes
Photos	Third-party apps	Yes	Yes
Messages/SMS/MMS	Date and time	Yes	Yes
Messages/SMS/MMS	Contact number	Yes	Yes
Call Logs	Incoming call	Yes	Yes
Call Logs	Outgoing call	Yes	Yes
Call Logs	Missed calls	Yes	Yes
Call Logs	Date and time	Yes	Yes
Notes	Search string	Yes	No
Notes	Date and time	Yes	No
Voice Memos	Date and time	Yes	No
Web History	Date and time	Yes	No
Email	Date and time	Yes	No
Facebook Messages	Date and time	*	No
WhatsApp Messages	Date and time	Yes	No
LinkedIn Messages	Date and time	*	No
WeChat Messages	Date and time	*	No
Viber Messages	Date and time	*	No

4.3 Off-Device Backup-Based Filtering

Table 2 shows the data that can and cannot be extracted by the TDES app in the off-device mode via metadata filtering. The first part of the table shows the data that can be extracted from iPhones and Android phones. The second part shows the data that can be extracted from Android phones, but not from iPhones. The third part shows data that the TDES app currently cannot extract from iPhones and Android phones. Note that a table entry marked with an asterisk (*) corresponds to an item that was not investigated.

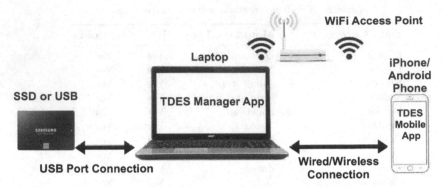

Figure 3.   TDES communications paradigm.

- **iPhones:** Apple iOS security mechanisms do not permit applications that execute on an iPhone to extract certain types of content (second and third sections of Table 1). Therefore, this content is acquired from an iTunes backup. The `idevicebackup2` command supported by the open-source `libimobiledevice` [21] is employed. Other standard, albeit complex, techniques can also be used to extract data from a backup.

- **Android Phones:** In the case of Android phones, any data that can be extracted off-device can also be extracted on-device; therefore, on-device extraction is employed. However, data from the third-party applications in Table 1 cannot be extracted using on-device acquisition when the phone is not rooted. Experiments with rooted and non-rooted Android phones did not reveal an Android equivalent of the iTunes backup mechanism.

4.4   TDES Communications

Communications between the TDES manager and the TDES app on a target phone is an important component of the TDES system. Figure 3 shows the communications paradigm that is implemented on iPhones and Android phones. The forensic investigator is provided with a portable TDES boot drive (e.g., SSD drive or USB stick) that is preloaded with a bootloader for a Windows 10 machine, TDES manager and the tools necessary to install the TDES app on the target phone. All the extracted data is sent back to the boot drive by the TDES app; reports pertaining to the extracted data also reside on the boot drive.

Any available Windows 10 system can be used to boot into the TDES manager, which runs in an isolated environment on the drive. After booting up, the TDES manager must have Internet access if the target device is an iPhone.

The steps for targeted data extraction are:

- The boot drive containing the TDES manager is inserted into a laptop.

- The Windows 10 operating system boots up and the TDES manager starts its execution.

- A wired connection using a USB cable is established from the laptop to the phone. The TDES app is installed. In the case of an iPhone, a hotspot is needed to connect to Apple in order to sign the code and acknowledge trust in the developer.

- After the app is downloaded, the phone may be disconnected from the laptop.

- A wireless or wired two-way communications channel is set up between the TDES manager and TDES app for data transfer.

- The targeted data extracted by the TDES app is exported to the TDES manager and reports are generated for the extracted data.

Note that no copies of data or residual data from the export process are stored on the phone.

- **TDES App Installation on iPhones:** Only applications from sources approved by Apple can be executed on iPhones that are not jailbroken. Apple iOS requires that all executable code must be signed with a certificate issued by Apple. Third-party apps must have signed certificates to ensure that they do not load any tampered or self-modifying code [6].

  The TDES implementation uses Cydia Impactor [34] to sign the TDES app code. The procedure involves the generation of an iOS App Store Package (IPA) file of the TDES app using the XCode Archive utility. This application archive file stores an iPhone app. In order to sign the code, Impactor logs into the Apple Developer Center and downloads the developer's provisioning profile and iOS development certificate. Logging into the Apple Developer Center requires an Internet connection. Impactor signs the IPA file content in a depth-first manner starting with the deepest folder level. After the signing is done, Impactor installs the TDES app on the iPhone. All these tasks are automated by an AutoHotKey script [7] that executes after the TDES manager boots; thus, no actions are required to be performed by the forensic investigator.

- **TDES App Installation on Android Phones:** The Android operating system permits only signed applications to be installed on an Android phone. As long as an application is signed and does not attempt to update another application, it can be self-signed – this approach is adopted in the TDES implementation. The output of the compilation is an APK file. Note that no other authentication is necessary.

  The TDES app is installed after the APK file is stored on the target phone. For simplicity and ease of use, an Android debug bridge is employed for communications between the host computer and target phone. The Android debug bridge requires the phone to be placed in the USB debugging mode; this mode is turned off after the app is installed.

- **TDES Data Transfer Protocol:** The communications channel between the TDES app and TDES manager must ensure that the extracted data is transmitted with forensic integrity and that all data modifications are detected and documented. Furthermore, data that is modified inadvertently or intentionally during the chain of custody is also identified and documented.

  This is implemented by hashing essentially every file and computing a final hash value, which is exported to the TDES manager. Note that the hashing is done on the phone. If required, the final hash value could be sent to the phone's owner, the forensic investigator or to a third party.

  The iPhone implementation employs a socket-based data transfer protocol. Since the iPhone implementation requires a hotspot in any case, a wireless link is used for communications between the app and the manager.

  The Android implementation uses an Android debug bridge, which supports socket-level communications. Since Android applications are natively written in Java, ServerSockets and Sockets are employed. A wired connection is used for the Android communications protocol.

4.5　　　User Interface

The user interface, which runs as part of the app on the target phone, enables a forensic investigator to specify the selection criteria for data extraction. At this time, the interfaces are somewhat different for iPhones and Android phones. An optional PDF consent form is provided by the TDES manager. In the case of an iPhone, after the data extraction criteria are specified using the app, a digital consent form that specifies the data to be extracted can be completed on the app itself. In the case of an Android phone, a broad consent form is completed on the app first. This consent form ensures that only the relevant subset of data specified using the app is, in fact, extracted.

A useful bookmarking feature is provided by the TDES app. Consider a situation where a dataset has been extracted using a set of filters. The forensic investigator who set up the filters can display the results and do a quick data review on the phone itself before deciding what data to actually export to the TDES manager (i.e., bookmarked data). For example, if the investigator selected a set of images of weapons obtained during a certain time period, then he/she could review the images and select a subset of relevant images by bookmarking the subset.

Discussions with a former prosecutor and a current defense attorney indicated that bookmarking is a useful feature, but it may introduce bias during the evidence collection process. Consequently, the current implementation enables bookmarking to be turned on or off. Alternatively, both options may be selected, producing two versions of the exported data – the bookmarked version and the original version. If needed, an investigator could export all the data that could be examined under the consent and filtering definitions, including possible exculpatory data.

- **iPhone App Interface:** Figure 4 shows the iPhone TDES app interface. The initial choices for a forensic investigator to define are: (i) when (specific date ranges, today, last week, last month, etc.); (ii) where (current location, location within a certain number of miles, location determined by city, state or zip code, etc.); and (iii) what (data types – photos, videos, calendar, call logs, messages and contacts).

  Additional filtering options – generally, content filtering – may be defined. For example, if photos and videos are of interest, then the content filtering options supported are the inclusion or exclusion of weapons, places, vehicles, drugs, websites, gadgets, skin exposure, pornography and favorites. If the exclude skin exposure option is selected, then the app filters the corresponding images, and displays and exports the remaining images.

  The last screen of the interface enables the investigator to display the selected data on the device, export the data, or both. A consent form is displayed before the data is exported to the TDES manager.

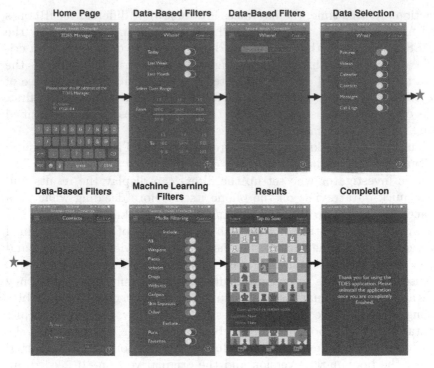

Figure 4.   iPhone TDES app user interface.

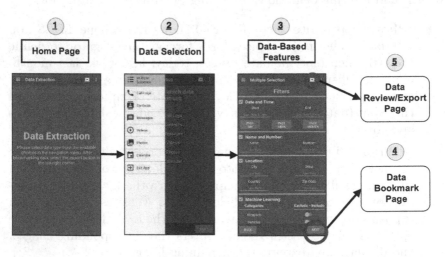

Figure 5.   Android TDES app user interface.

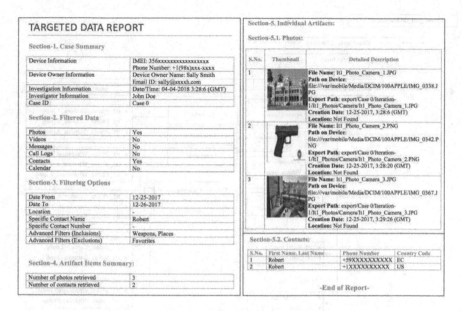

Figure 6. TDES summary report for an iPhone.

- **Android App Interface:** Figure 5 shows the Android TDES app interface. The app first presents a screen for specifying the data categories to be extracted; the same categories of data as the iPhone app are supported. Selecting any of these data types leads to a new screen with another set of choices providing additional filtering options for metadata and content filtering. The Android app interface also has provisions for first defining a broad consent form that restricts further data selections. It also supports data bookmarking, display and export.

Both versions of the app interface support a fair amount of metadata and content filtering. For example, call logs can be filtered by name and number as well as by date and time. Contacts can be filtered by name and number. Messages can be filtered by name and number as well as by date and time. Videos and photos can be filtered by location, date, time and various implemented content using machine learning models.

4.6 Reporting and Forensic Integrity

A common interface using the JSON object format [11] is implemented for the selected export of data from the iPhone and Android phone apps. The JSON structure facilitates the description of the extracted data as well as hash values and reporting information. For example, a report

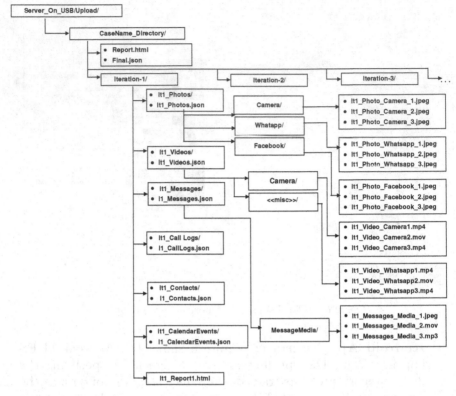

Figure 7.   Output file structure.

may need to document when the TDES app began its execution and when the extraction was completed. Although the data transfer is primarily from the app to the manager, some information, such as the forensic investigator's name, phone owner's name and case number, is passed from the manager to the app. The Android TDES app extracts additional information such as the IMEI, phone number and email address associated with the phone. In the case of the iPhone TDES app, this information must be entered in the manager. Figure 6 shows a sample report generated for an iPhone.

- **TDES Directory Structure on the Boot Drive:** Figure 7 shows the directory structure created for storing evidence on the boot drive. The structure is designed to ensure data integrity and support reporting. A directory is created for each case. The complete report is stored as an HTML file in this directory. The JSON files, including `Final.json`, are discussed below. The extracted

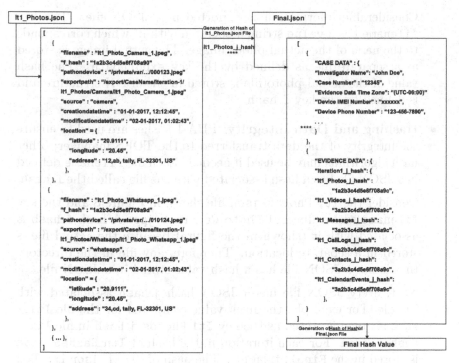

Figure 8. Example JSON files.

data is stored as one or more iterations of requests made by the investigator. In each iteration, every data category has a separate directory and a JSON file is associated with the directory.

■ **JSON Format for Data Transfer:** The JSON format is used to describe the structure of the exported data, which is used to create reports in the HTML format. Figure 8 shows example JSON files.

Assume that a set of photos has been extracted using metadata and content filters. Auxiliary information about each photo is transferred to the TDES manager along with the actual image file. The TDES apps for iPhones and Android phones create this information in the same format. After the information is transferred to the TDES manager, a report manager creates the actual report. Hashes are also transferred as part of the JSON files. As shown in Figure 8, the It1_Photos.json file is structured into arrays of arrays containing (key, value) pairs. For example, creation_date is a key and its value is the string 01-01-2017.

Considerable information is exported in a JSON file. The key filename has a value string associated with it, which corresponds to the name of the actual photo image. The actual image is stored as a separate file as defined by the key exportpath. The hash value of the actual photo file is stored in the JSON structure and is defined by the key f_hash.

- **Hashing and Data Integrity:** SHA-1 hashes are used to ensure the integrity of the data transferred to the TDES manager; other hash algorithms may be used if needed. Each file filename defined in a JSON file has a hash associated with the file called the f_hash.

  Consider the It1_Photos.json file shown in Figure 8 and the key filename with value It1_Photo_Camera_1.jpg. A hash f_hash is associated with it (shown in the figure) because the actual file is stored in a separate location. Therefore, any file in the directory that is not a JSON file has a hash value stored in a JSON file.

  Next, every JSON file has a JSON hash j_hash associated with the file. For example, the hash value computed for the file It1_-Photos.json is stored as the key It1_Photos_j_hash in file It1_-Hashes.json. For each iteration n, the hash of Itn_Hashes.json is stored in the Final.json file. The hash of Final.json is called Final_hash. This hash value ensures that no file in any case directory can be modified without detection.

  The Final_hash value computed by an app is sent to the TDES manager and stored in Report.htm. The manager can independently compute the Final_hash value to check if any changes occurred during the data transfer. Hash values are computed at intermediate points for several reasons, including to facilitate the granular transfer of data and check if the transfer is correct. Checking the extracted files against known files is also simplified. The TDES manager (or app) could also email a copy of the Final_hash value to the phone's owner, forensic investigator or third party.

5.   Experiments and Results

Several experiments were conducted to evaluate the accuracy and speed of selective data filtering on iPhone and Android phones. The metadata filtering accuracy should be 100% because the Apple and Android frameworks were employed; however, manual checks of metadata filtering were still performed.

The performance of the prototype system was also compared against two commercial tools, Paraben EEE and Magnet AXIOM, which are

Table 3. Devices used in the experiments and device content.

Model/Version	NIC	P	V	M	CL	CO	CA
Device 1 iPhone-8 (iOS v11.2.1)	Lightning port	10,307	178	208	482	1,102	148
Device 2 iPhone-7 (iOS v11.2.5)	Lightning port	2,621	109	5	155	6	46
Device 3 iPhone-6 Plus (iOS v11.2.2)	Lightning port	2,566	102	15,978	714	384	265
Device 4 Samsung Galaxy S7 (v7.0, Nougat)	Micro-USB 2.0	100	6	37	7	20	17
Device 5 Moto G3 (v6.0, Marshmallow)	Micro-USB 2.0	191	7	25,420	429	1,889	780
Device 6 Samsung Galaxy S7 Edge (v7.0, Nougat)	Micro-USB 2.0	249	22	13,362	500	240	337

**NIC: Network Interface Card; P: Photos; V: Videos;
M: Messages; CL: Call Logs; CO: Contacts; CA: Calendar**

used by law enforcement. As mentioned above, neither of these tools (nor Cellebrite) can perform selective data extraction as implemented by the prototype system. Note that the Cellebrite commercial tool was not evaluated because this tool (like the others) essentially performs a physical acquisition of all the phone data and then enables the user to analyze the data off-device.

Three iPhones and three Android phones were used in the experiments. Table 3 provides details about the phones and their contents. Apple Devices 1 and 3, which belong to the authors of this chapter, contained real user data. Apple Device 2 contained synthetic, noncopyrighted data that is available for reuse over the Internet. Similarly, Android Devices 5 and 6 contained real user data and belong to the authors; Apple Device 4 contained synthetic data. Table 3 also shows the total numbers of artifacts of each data category residing in each test device. The TDES boot drive used was a SanDisk Extreme 128 GB

Table 4.   On-device metadata-based filtering for iPhones.

| Category: Filter | Device 1 Experiments | | | |
	Artifacts	Display Time	Export Time	Size
1-Photos: 12/24/17–12/27/17	2/10,307	0.7 s	3.58 s	2.33 MB
2-Photos: Within 10 miles*	418/10,307	1.21 s	42 m, 64 s	822 MB
3-Videos: 09/1/17–01/31/18	34/178	1.20 s	51 m, 11 s	1,038 MB
4-Videos: Within 10 miles*	–	–	–	–
5-Videos: Current location*	4/178	0.2 s	17 m, 2 s	405 MB
6-Contacts: "Puppy"	3/1,102	2.57 s	0.6 ms	–
7-Contacts: "Robert"	–	–	–	–
8-Contacts: (xxx)xxx–xxx	1/1,102	0.12 s	0.8 ms	–
9-Calendar: 01/01/18–01/15/18	19/148	0.14 s	0.6 ms	–
10-Photos: 08/30/17–09/15/17	91/10,307	0.7 s	4 m, 1 s	236 MB
Videos: Any location	1/178			
11-Photos: 08/31/17	9/10,307	0.73 s	1 m, 1 s	51 MB
Videos: Within 50 miles	1/178			
12-Videos: Last week	3/178	0.4 s	1 m, 2 s	47 MB
Within 10 miles				

stick. A ThinkPad X1 Carbon laptop was used as the boot drive and to connect to the test phones.

iPhone Results.   The iPhone experiments employed Devices 1, 2 and 3. Table 4 shows the results for on-device metadata-based filtering for Device 1. Each experiment (row) focuses on a specific data category and filter. For each experiment, the total number of artifacts selected out of the total number of artifacts on the device is shown (e.g., in the case of the 1-Photos experiment, 2/10,307 means that two photos out of 10,307 photos on the device were extracted). The metadata filtering was 100% accurate based on manual checking (e.g., a phone feature such as Photos Album count). The table also shows the times required to display data on the target device and to export data to the TDES manager (via a wired connection). The recorded times show that TDES is feasible for in-field targeted data extraction. The amounts of exported data are also shown. Note that a table entry marked with an asterisk (*) corresponds to an item whose location depends on the physical location of the phone.

Table 5 shows the results of experiments for off-device backup-based metadata filtering for Device 3. The results for messages and call logs are

Table 5. Off-device backup-based filtering for iPhones.

Device 3 Experiments		
Category: Filter	Artifacts	Display Time
1-Messages: None	15,978/15,978	1.95 s
2-Messages: 10/03/17–12/30/17	510/15,978	0.33 s
3-Messages: (***)***–***	1,016/15,978	0.29 s
4-Call Logs: None	683/683	0.29 s
5-Call Logs: 01/14/17–08/14/17	297/683	0.27 s
6-Call Logs: (***)***–***	40/683	0.27 s
7-Messages: 01/14/17–08/14/17	738/15,978	0.32 s
Call Logs: (***)***–***	35/683	
8-Messages: (***)***–***	1,016/15,978	0.28 s
Call Logs:	40/683	

shown. As discussed earlier, the backup-based procedure involved the TDES manager acquiring a complete backup from iTunes; thus, there is no export time. Note, however, that the forensic investigator must still specify the filtering that must be performed by the TDES app. The accuracy of metadata filtering is always 100% based on manual analysis using iTunes.

Table 6. On-device metadata and content filtering for iPhones (Inception-v3).

Device 2 Experiments				
Category: Filter	Content Filter	Display Time	Export Time	Accuracy (%)
1-Photos: 12/25/17	Weapons	9.82 s	4.69 s	97.22
2-Photos: Within 10 miles	Weapons	20.31 s	–	–
3-Photos: 12/25/17–12/29/17	Weapons	19.05 s	4.34 s	94.50
4-Photos: 12/25/17–12/29/17	Places	17.73 s	9.93 s	88.07
5-Photos: 12/25/17–12/29/17	Vehicles	17.06 s	0.72 s	100.00
6-Photos: 12/25/17–12/29/17	Drugs	16.26 s	0.33 s	96.33
7-Photos: 12/25/17–12/29/17	Websites	16.66 s	7.03 s	99.08
8-Photos: 12/25/17–12/29/17	Gadgets	17.33 s	7.88 s	89.91
9-Photos: 12/25/17–12/29/17	Skin exposure	16.23 s	8.68 s	100.00

Table 6 shows the results of the experiments using Device 2 photos for various combinations of metadata and content filtering. The test

Table 7.   On-device metadata-based filtering for Android phones.

Category: Filter	Device 5 Experiments Artifacts	Display Time	Export Time	Size
1-Photos: 02/03/18–02/05/18	2/191	0.31 s	2.32 s	6.56 MB
2-Photos: Current location	1/191	0.63 s	10.58 s	46.1 MB
3-Videos: 12/19/17–02/03/18	3/7	0.89 s	3.91 s	16.6 MB
4-Videos: Current location	7/7	0.90 s	13.09 s	190 MB
5-Calendar: 05/29/17–05/30/17	85/780	1.03 s	2.40 s	13 KB
6-Messages: "aaabb"	32/25,420	1.23 s	14.23 s	7 KB
7-Messages: (***)***_***	5/25,420	0.92 s	1.25 s	4 KB
8-Call Logs: "aaabb"	9/429	0.49 s	6.59 s	5 KB
9-Call Logs: (***)***_***	11/429	0.89 s	11.2 s	6 KB
10-Messages: (***)***_***	100/25,420	1.25 s	14.08 s	199.1 MB
Photos: 01/28/18–02/05/18	6/191			
Videos: Current location	7/7			
11-Messages: 12/12/17–02/05/18	1,000/25,420	1.02 s	3.89 s	258 KB
Call Logs: (***)***_***	8/429			
12-Messages: 09/12/17–09/29/17	300/25,420	1.65 s	18.2 s	236.1 MB
Calendar: 09/12/17–09/29/17	5/780			
Photos: Current location	1/191			
Videos: Current location	7/7			

iPhone had 2,621 photos with 109 photos in the date range 12/25/17 to 12/29/17, and 72 of these photos were taken on 12/25/17. The Inception-v3 model was used for content filtering. Rows 1–3 of the table focus on filtering for "weapons." In the case of Row 2, content filtering was not applied because none of the weapons photos were taken within 10 miles. Rows 4–9 focus on content filters that would be relevant to law enforcement. The times required for display and export are shown for each experiment. The accuracy measure expresses how well the Inception-v3 model performs content filtering. The accuracy computations involved the creation of a confusion matrix for each experiment, following which the accuracy was computed as:

$$Accuracy = \frac{TP + TN}{TP + FN + FP + TN} \times 100 \qquad (1)$$

where TP denotes true positive; TN denotes true negative; FP denotes false positive; and FN denotes false negative.

Table 8. On-device metadata and content filtering for Android phones (MobileNet).

| Category: Filter | Device 4 Experiments | | | |
	Content Filter	Display Time	Export Time	Accuracy (%)
1-Photos: 11/12/17–02/02/18	Weapons	35 s	1.5 s	75.68
2-Photos: Current location	Weapons	1.3 s	0.81 s	100.00
3-Photos: 10/12/17–12/02/17	Vehicles	37.4 s	1.56 s	25.00
4-Photos: Current location	Vehicles	1.4 s	1.3 s	100.00
5-Photos: 12/01/17–01/13/18	Drugs	34.69 s	1.2 s	92.06
6-Photos: Current location	Drugs	1.2 s	0.0 s	71.43
7-Photos: 08/11/17–12/31/17	Skin exposure	33.08 s	2.48 s	92.21

Android Phone Results. The Android phone experiments employed Devices 4, 5 and 6. Table 7 shows the results for on-device metadata-based filtering for Device 5. Each experiment (row) focuses on a specific data category and filter. Note that the display and export times are very good. For example, in the case of the 12-Messages experiment, exporting 236 MB of device artifacts required only 18.2 seconds.

Table 8 shows the results of seven experiments using Device 4 photos for various combinations of metadata and content filtering. The MobileNet model from TensorFlowLite was used for metadata and content filtering. The display and export time results are excellent. The accuracy measure, computed using Equation (1), expresses how well the MobileNet model performs content filtering. The results are modest; better machine learning models will have to be developed to improve the accuracy of content filtering.

Comparison with Commercial Tools. Several experiments were conducted to compare the data export times for TDES against the times required by two commercial tools, Paraben and Magnet AXIOM. iPhone Device 2 and Android Device 4 were used in the experiments. Table 9 shows the experimental results – the iPhone comparisons are in the top half of the table and the Android comparisons are in the bottom half of the table. The app installation time (AIT) is the time period from the instant the target device was connected to the laptop to the time when a data selection can be made (in the case of TDES, this is when a data selection can be made on the target device; in the case of Paraben and Magnet AXIOM, this is when a data selection choice can be made on the laptop). The backup acquisition time (BAT) is the time taken for backup-based acquisition. Note that, in the case of TDES, the exported

Table 9.   Export time comparisons for iPhone Device 2 and Android Device 6.

Item	TDES	Paraben	Magnet AXIOM
Device 2 (iPhone)			
AIT	52 s	10 m	9 m
BAT	26 m (2 GB)	20 m	38 m, 54 s (4.1 GB)
Call Logs	(BAT) 15 ms	0.1 s	0.4 s
Messages	(BAT) 16 ms	0.1 s	0.3 s
Contacts	1.8 ms	0.2 s	0.3 s
Calendar	2 ms	0.2 s	0.3 s
Photos	39 m, 3 s (2,621 files)	–	80 m (29,488 files, 2.30 GB)
Videos	30 m, 15 s (109 files)	–	4 m (438 files, 1.73 GB)
All Media	Not needed	32 m	93 m (48,701 files, 2.69 GB)
Device 6 (Android)			
AIT	14 s	5 s	NA
BAT	NA	NA	29 m
Call Logs	1 s	40 s	1 m, 17 s
Messages	4 m, 9 s	17 m, 3 s	1 m, 21 s
Contacts	1 s	2 m, 11 s	1 m, 11 s
Calendar	6 s	1 m, 5 s	1 m, 14 s
Photos	42 s (249 files)	–	14 m, 41 s (13,711 files)
Videos	14 s (22 files)	–	1 m, 38 s (62 files)
All Media	NA	43 s	NA

data was stored on a flash drive whereas, in the case of Paraben and Magnet AXIOM, the exported data was stored on the laptop hard drive.

- **iPhone Comparison:** The installation time of the TDES app on the iPhone was 52 seconds. Paraben and Magnet AXIOM had to first create a backup of the iPhone data. In the case of Paraben, backup creation (20 minutes) occurs in conjunction with application initialization (10 minutes) whereas Magnet AXIOM has a separate backup creation step of 38 minutes and 54 seconds after 9 minutes of application initialization. Note that TDES has a backup acquisition time only when extracting call logs and messages.

- **Android Phone Comparison:** The installation time of the TDES app on the Android phone was 14 seconds. Since Magnet AXIOM uses backup-based acquisition, a backup must be created before extracting any artifacts. For example, when extracting call logs, Magnet AXIOM created a backup that took 29 minutes followed by call log extraction that took one minute and 17 seconds. The TDES app required 14 seconds for app installation and one second for data export. In contrast, Paraben required five seconds for initialization and 40 seconds for data export.

In the case of Paraben and Magnet AXIOM, the only choices available for acquisition are the broad categories shown in Table 9. Paraben does not extract photos and videos separately; it provides one option for all media artifacts. However, experiments revealed that selecting this option resulted in the extraction of metadata associated with media artifacts, not the artifacts themselves.

6. Conclusions

The targeted data extraction system described in this chapter supports the acquisition of relevant data from iOS and Android devices in a forensically-sound manner. It implements state-of-the-art metadata and content filtering functionality based on machine learning techniques. Forensic soundness is realized using the eDiscovery Reference Model [19] and dynamic/live analysis techniques drawn from network and cloud forensics [17, 26]. The design assumes that a phone is voluntarily provided to law enforcement under a documented consent agreement. However, it is equally applicable to situations where a court orders that a smartphone passcode must be provided for evidence recovery or where a smartphone memory dump (e.g., from a cloud backup) with an intact filesystem is available. The targeted data extraction system is currently being provided to law enforcement for testing and feedback, with the goal of incorporating additional features and capabilities.

Acknowledgements

This research was supported in part by the National Institute of Justice, Office of Justice Programs, U.S. Department of Justice under Award No. 2016-MU-CX-K003. The opinions, findings and conclusions or recommendations expressed in this chapter are those of the authors and do not necessarily reflect the opinions of the U.S. Department of Justice.

References

[1] M. Abadi, P. Barham, J. Chen, Z. Chen, A. Davis, J. Dean, M. Devin, S. Ghemawat, G. Irving, M. Isard, M. Kudlur, J. Levenberg, R. Monga, S. Moore, D. Murray, B. Steiner, P. Tucker, V. Vasudevan, P. Warden, M. Wicke, Y. Yu and X. Zheng, Tensorflow: A system for large-scale machine learning, *Proceedings of the Twelfth*

USENIX Symposium on Operating Systems Design and Implementation, pp. 265–283, 2016.

[2] N. Al Mutawa, I. Baggili and A. Marrington, Forensic analysis of social networking applications on mobile devices, *Digital Investigation*, vol. 9(S), pp. S24–S33, 2012.

[3] A. Aminnezhad, A. Dehghantanha and M. Abdullah, A survey of privacy issues in digital forensics, *International Journal of Cyber-Security and Digital Forensics*, vol. 1(4), pp. 311–323, 2012.

[4] C. Anglano, Forensic analysis of WhatsApp Messenger on Android smartphones, *Digital Investigation*, vol. 11(3), pp. 201–213, 2014.

[5] Apple, Core ML, Cupertino, California (`developer.apple.com/documentation/coreml`), 2017.

[6] Apple, iOS Security, iOS 12.3, Cupertino, California (`www.apple.com/business/docs/iOS_Security_Guide.pdf`), 2019.

[7] AutoHotkey Foundation, AutoHotkey (`autohotkey.com`), 2019.

[8] J. Bergstra, F. Bastien, O. Breuleux, P. Lamblin, R. Pascanu, O. Delalleau, G. Desjardins, D. Warde-Farley, I. Goodfellow, A. Bergeron and Y. Bengio, Theano: Deep learning on GPUs with Python, *Proceedings of the BigLearning Workshop*, vol. 3, 2011.

[9] G. Cantrell, D. Dampier, Y. Dandass, N. Niu and C. Bogen, Research toward a partially-automated and crime-specific digital triage process model, *Computer and Information Science*, vol. 5(2), pp. 29–38, 2012.

[10] K. Cho, B. van Merrienboer, C. Gulcehre, D. Bahdanau, F. Bougares, H. Schwenk and Y. Bengio, Learning Phrase Representations using RNN Encoder-Decoder for Statistical Machine Translation, arXiv:1406.1078 (`arxiv.org/abs/1406.1078`), 2014.

[11] D. Crockford, The application/json Media Type for JavaScript Object Notation (JSON), RFC 4627, 2006.

[12] Google, Android Developer Manual, Mountain View, California, 2017.

[13] A. Howard, M. Zhu, B. Chen, D. Kalenichenko, W. Wang, T. Weyand, M. Andreetto and H. Adam, MobileNets: Efficient Convolutional Neural Networks for Mobile Vision Applications, arXiv:1704.04861 (`arxiv.org/abs/1704.04861`), 2017.

[14] M. Husain, I. Baggili and R. Sridhar, A simple cost-effective framework for iPhone forensic analysis, *Proceedings of the International Conference on Digital Forensics and Cyber Crime*, pp. 27–37, 2010.

[15] Y. Jia, E. Shelhamer, J. Donahue, S. Karayev, J. Long, R. Girshick, S. Guadarrama and T. Darrell, Caffe: Convolutional architecture for fast feature embedding, *Proceedings of the Twenty-Second ACM International Conference on Multimedia*, pp. 675–678, 2014.

[16] Keras Team, Keras: Deep Learning for Humans, GitHub (`github.com/keras-team/keras`), 2019.

[17] S. Khan, A. Gani, A. Abdul Wahab, M. Shiraz and I. Ahmad, Network forensics: Review, taxonomy and open challenges, *Journal of Network and Computer Applications*, pp. 214-235, 2016.

[18] A. Krizhevsky, I. Sutskever and G. Hinton, ImageNet classification with deep convolutional neural networks, in *Communications of the ACM*, vol. 60(6), pp. 84–90, 2017.

[19] D. Lawton, R. Stacey and G. Dodd, E-Discovery in Digital Forensic Investigations, CAST Publication Number 32/14, Centre for Applied Science and Technology, Home Office, London, United Kingdom, 2014.

[20] Y. LeCun, Y. Bengio and G. Hinton, Deep learning, *Nature*, vol. 521(7553), pp. 436–444, 2015.

[21] `libimobile.org`, `libimobiledevice`: A cross-platform software protocol library and tools to communicate with iOS devices natively (`www.libimobiledevice.org`), 2019.

[22] J. Mahadeokar, Open NSFW Model, GitHub (`github.com/yahoo/open_nsfw`), 2017.

[23] S. Morrissey and T. Campbell, *iOS Forensic Analysis: For iPhone, iPad and iPod touch*, Apress, New York, 2010.

[24] K. Murphy, *Machine Learning: A Probabilistic Perspective*, MIT Press, Cambridge Massachusetts, 2012.

[25] D. Quick and M. Alzaabi, Forensic analysis of the Android filesystem YAFFS2, *Proceedings of the Ninth Australian Digital Forensics Conference*, pp. 100–109, 2011.

[26] V. Roussev, A. Barreto and I. Ahmed, Forensic Acquisition of Cloud Drives, arXiv:1603.06542 (`arxiv.org/abs/1603.06542`), 2016.

[27] V. Roussev, C. Quates and R. Martell, Real-time digital forensics and triage, *Digital Investigation*, vol. 10(2), pp. 158–167, 2013.

[28] N. Scrivens and X. Lin, Android digital forensics: Data, extraction and analysis, *Proceedings of the ACM Turing 50th Celebration Conference – China*, article no. 26, 2017.

[29] A. Shekhar, Android TensorFlow Machine Learning Example, *MindOrks Blog* (`blog.mindorks.com/android-tensorflow-machine-learning-example-ff0e9b2654cc`) March 6, 2017.

[30] Y. Steinbuch and J. Tacopino, Woman records horrific scene after boyfriend is fatally shot by police, *New York Post*, July 7, 2016.

[31] P. Stirparo and I. Kounelis, The Mobileak Project: Forensic methodology for mobile application privacy assessment, *Proceedings of the International Conference on Internet Technology and Secured Transactions*, pp. 297–303, 2012.

[32] M. Stroud, In Boston bombing, flood of digital evidence is a blessing and a curse, *CNN*, April 18, 2013.

[33] C. Szegedy, V. Vanhoucke, S. Ioffe, J. Shlens and Z. Wojna, Rethinking the inception architecture for computer vision, *Proceedings of the IEEE Conference on Computer Vision and Pattern Recognition*, pp. 2818–2826, 2016.

[34] Team Cydia, Cydia Impactor (`cydia-app.com/cydia-impactor`), 2019.

[35] TensorFlow, Introduction to TensorFlow Lite (`www.tensorflow.org/mobile/lite`, 2018.

Chapter 6

EXPLOITING VENDOR-DEFINED MESSAGES IN THE USB POWER DELIVERY PROTOCOL

Gunnar Alendal, Stefan Axelsson and Geir Olav Dyrkolbotn

Abstract   The USB Power Delivery protocol enables USB-connected devices to negotiate power delivery and exchange data over a single connection such as a USB Type-C cable. The protocol incorporates standard commands; however, it also enables vendors to add non-standard commands called vendor-defined messages. These messages are similar to the vendor-specific commands in the SCSI protocol, which enable vendors to specify undocumented commands to implement functionality that meets their needs. Such commands can be employed to enable firmware updates, memory dumps and even backdoors.

This chapter analyzes vendor-defined message support in devices that employ the USB Power Delivery protocol, the ultimate goal being to identify messages that could be leveraged in digital forensic investigations to acquire data stored in the devices.

Keywords: USB Power Delivery protocol, vendor-specified messages, exploitation

1.   Introduction

An important goal of mobile device forensics is to acquire data. Mobile phones typically have two key data sources: (i) volatile memory (RAM); and (ii) long-term storage (typically, flash memory). These two sources differ in content and acquisition methods. RAM is often proprietary, short-term storage that is not intended for interpretation by applications other than the one that stored the data. In contrast, long-term storage such as flash memory contains well-structured data, usually in a filesystem, that is meant to be re-read, typically by the operating system. Nevertheless, both types of storage maintain data that is important in digital forensic investigations.

© IFIP International Federation for Information Processing 2019
Published by Springer Nature Switzerland AG 2019
G. Peterson and S. Shenoi (Eds.): Advances in Digital Forensics XV, IFIP AICT 569, pp. 101–118, 2019.
https://doi.org/10.1007/978-3-030-28752-8_6

Security mechanisms in commercial products are hindering the forensic acquisition of data. Data encryption in flash memory has invalidated methods such as desoldering (i.e., chip-off) that enable data to be read directly from a chip. Encryption prevents the extracted data from being interpreted without the decryption keys. The keys are often protected by additional encryption keys that tie the data to the specific device that encrypted the data in long-term storage. Therefore, transplanting a flash memory chip to a different, but identical, device would not decrypt the stored data. Device-tied encryption keys are also protected by security features such as TrustZone that rely on tamper-proof hardware. Therefore, in order to access data from a secured device, it is necessary to exploit security vulnerabilities in the device itself, or leverage undocumented features such as backdoors or indirectly increase the attack surface of the device.

The general approach is that any data extraction technique should be researched extensively, including any and all means it uses to communicate with other devices. The USB Power Delivery protocol is a communications mode that has the potential to increase the device attack surface. The idea is that, if undocumented means exist to communicate with the device, then hidden features and security vulnerabilities could be identified and exploited to facilitate data acquisition.

The USB Power Delivery protocol provides a uniform means for vendors to implement power negotiation between power sources and devices such as mobile phones and personal computers in order to maximize the charging current. The power source can support different power configurations, one power profile for a mobile phone and a different profile for a personal computer, to enable the devices to obtain the appropriate currents and voltages. Devices can also use the protocol to request higher currents and voltages from power sources. In the case of two non-power-source devices (e.g., two mobile phones), the devices can negotiate a power delivery profile so that one device can charge the other. Another example is a monitor connected to a personal computer where the protocol enables the monitor to draw power from the personal computer if it is not connected to an external power source. If the monitor is connected to an external power source, then it could provide power to the personal computer. All these negotiations occur over the same USB cable unbeknownst to the user.

The USB Power Delivery protocol is of interest from a digital forensics perspective because it supports inter-device communications. These communications could be exploited to expand the attack surface of one or both devices, enabling the acquisition of data that is otherwise inaccessible. The focus is on vendor-defined messages in the USB Power

Delivery protocol. Undocumented messages discovered in other protocols have been demonstrated to enable firmware updates, memory dumps and even backdoors. This chapter presents a black-box testing approach for revealing proprietary messages supported by the USB Power Delivery protocol that could be leveraged in digital forensic investigations to acquire data stored in devices that support the protocol.

2. Related Work

Allowing vendors to incorporate proprietary vendor-defined messages or commands in protocols to provide custom functionality has led to the release of numerous consumer devices that potentially respond to undocumented commands with unknown behavior. This can have devastating security implications. As demonstrated by Alendal et al. [2], vendor-specified SCSI commands can be used to bypass authentication on self-encrypting hard drives. Whether this research represents the best-case scenario for law enforcement or the worst-case scenario for the vendor, one cannot ignore the fact that the existence of hidden commands must be tested carefully. Indeed, as devices and firmware change over time, such testing should be performed regularly by law enforcement and security researchers.

Testing the USB Power Delivery protocol for hidden commands requires a means for emulating the protocol. Reydarns et al. [5] have demonstrated the use of USB Power Delivery protocol emulation in testing different power configurations for a power source. However, there is little, if any, research on the security of the USB Power Delivery protocol and nothing related to digital forensics. This research is important because it comprehensively analyzes the USB Power Delivery protocol and attempts to discover how vendor-defined protocol messages could be leveraged to assist digital forensic examinations of devices that support the protocol.

3. USB Power Delivery Protocol

Revision 1.0 (version 1.0) of the USB Power Delivery protocol specification was released in 2012; several revisions have been released since, the most recent being Revision 2.0 (version 1.3) and Revision 3.0 (version 1.2) [8]. The protocol provides a uniform means for devices to negotiate power supply configurations across vendors. It is typically supported by devices with a USB Type-C port/connector with dedicated CC1 and CC2 links (Figure 1). The USB Type-C connection is reversible, enabling devices to communicate on either CC line.

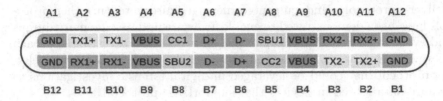

Figure 1. USB Type-C pinout [4].

The message-based USB Power Delivery protocol has three types of messages: (i) control messages; (ii) data messages; and (iii) extended messages. Control messages are short messages that typically require no data exchange. Data messages contain data objects that are transmitted between devices. Extended messages are essentially data messages with larger data payloads. The USB Power Delivery protocol leverages the three message types to define a wide range of standard messages, which enable devices to communicate and negotiate power source configurations.

Preamble	SOP Start of Packet	Message Header 16 bit	Data Objects (0-7) 32 bit	CRC	EOP End of Packet

Figure 2. Data message packet.

Figure 2 shows a data message packet comprising a preamble for synchronization, start of packet (SOP), message header, up to eight data objects of 32-bits each, CRC and end of packet (EOP). The preamble, SOP, CRC and EOP are part of the physical transport layer; they are common to all three types of messages, along with the message header. The optional data objects are only found in data messages.

Table 1 lists example control and data messages in the USB Power Delivery protocol.

The USB Power Delivery protocol supports different standard message sets as indicated by the protocol specification revisions, currently Revision 2.0 and Revision 3.0. Revision 3.0 is functionally the same as Revision 2.0, except for new features such as USB authentication. Interested readers are referred to the protocol specifications [8] for information pertaining to the differences between the message sets.

The USB Power Delivery protocol also enables cables to take part in communications; a device can communicate with a cable directly using the start of packet. Such electronically-marked cables (EMCA) enable devices to ensure that the cable supports high voltage/current power

Table 1. Control and data messages in Revision 3.0 (version 1.2).

Control Messages	Data Messages
GoodCRC	Source_Capabilities
GotoMin	Request
Accept	BIST
Reject	Sink_Capabilities
Ping	Battery_Status
PS_RDY	Alert
Get_Source_Cap	Get_Country_Info
Get_Sink_Cap	Vendor_Defined
DR_Swap	
PR_Swap	
VCONN_Swap	
Wait	
Soft_Reset	
Not_Supported	
Get_Source_Cap_Extended	
Get_Status	
FR_Swap	
Get_PPS_Status	
Get_Country_Codes	

source configurations. According to the protocol specification, devices can negotiate direct current levels up to 5 A, corresponding to a maximum of 100 W at 20 V between devices connected via an EMCA cable. Passive (non-EMCA) cables are rated for a maximum direct current of 3 A, which corresponds to 15 W at 5 V, 36 W at 12 V or 60 W at 20 V.

Figure 3 shows a typical power delivery negotiation – referred to as an explicit contract between two devices or port pairs. According to the standard, all port pairs are required to make an explicit contract. In a contract, the device (port) that consumes power is called the sink and the device (port) that provides power is called the source.

Vendors may implement novel functionality using proprietary vendor-defined messages, a subgroup of data messages in the USB Power Delivery protocol. Similar features are found in other protocols, such as vendor-specific commands in the SCSI protocol [6]. These commands are implemented and used only by vendors for internal purposes such as debugging, factory setup and proprietary communications with vendor software; the commands are not used in normal device operations. Vendor commands are rarely documented because they are reserved for internal use.

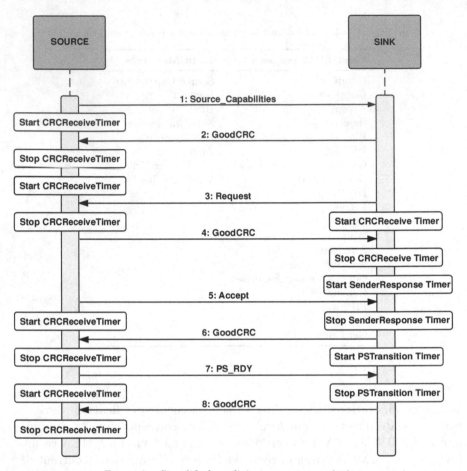

Figure 3. Simplified explicit contract negotiation.

Preamble	SOP Start of Packet	Message Header 16 bit	VDM Header (S)VID 16-bit | Command 16-bit	VDO (0-6) 32 bit	CRC	EOP End of Packet

Figure 4. Vendor-defined message packet.

Figure 4 shows a vendor-defined message (VDM) packet in the USB Power Delivery protocol. Vendor-defined messages are of two types: (i) structured; and (ii) unstructured. Structured vendor-defined message commands are defined in the USB Power Delivery protocol standard whereas unstructured vendor-defined message commands are implemented by vendors on an *ad hoc* basis. Note that a "command" is a subgroup of "message," which is either a structured vendor-defined mes-

SVID/VID Bit 31...16	VDM Type Bit 15	VDM Version Bit 14...13	*Reserved* Bit 12...11	Object Position Bit 10...8	Cmd Type Bit 7...6	*Reserved* Bit 5	Command Bit 4...0

Figure 5. Structured vendor-defined message header.

Vendor ID (VID) Bit 31...16	VDM Type Bit 15	Vendor Use Bit 14...0

Figure 6. Unstructured vendor-defined message header.

sage or an unstructured vendor-defined message. Thus, while structured vendor-defined messages have predefined command sets in the protocol specification, unstructured vendor-defined messages can correspond to commands defined by vendors.

Because vendor-defined messages are a type of data message, there is a size limitation on the amount of data a message can contain – this corresponds to the size of six vendor data objects (VDOs) plus the 32-bit vendor-defined message header. A vendor data object contains a 32-bit value (data). To prevent vendors from implementing conflicting messages, the protocol requires either the standard vendor ID (SVID) defined in the protocol specification or a vendor ID (VID) to be part of the vendor-defined message header. This means that a vendor must use one of its 16-bit USB Implementers Forum (USB-IF) vendor IDs [7] in all the vendor-defined messages it implements.

Example vendor IDs are 0x05ac (Apple) and 0x04e8 (Samsung). As shown in Figures 5 and 6, the structured vendor ID and vendor ID are required to be part of the corresponding vendor-defined message headers. Thus, a vendor with a valid USB-IF-assigned vendor ID can implement any command that contains up to six additional vendor data objects in one vendor-defined message. The command is the second part of the vendor-defined message header that can be any 15-bit value in the case of an unstructured vendor-defined message.

Table 2 shows example structured vendor-defined message commands.

4. Methodology

Devices come in different architectures from numerous vendors and without source code or firmware that implement the USB Power Delivery protocol. Therefore, a black-box method was attempted to test the existence of vendor-defined messages in the protocol. One approach is to analyze protocol communications between devices from the same vendor and determine if vendor-defined messages are employed. This

Table 2.   Structured commands in Revision 3.0 (version 1.2).

Structured Vendor-Defined Message Commands
Discover Identity
Discover SVIDs
Discover Modes
Enter Mode
Exit Mode
Attention
SVID Specific Commands (defined by the SVID)

assumes that, if such messages exist, the connected devices initiate their use by default.

Instead, a more active approach that directly communicates with a test device was employed. Since no solution was available to communicate with devices via the USB Power Delivery protocol, a home-grown approach was employed. A detailed description of this approach is beyond the scope of this chapter. However, the concept is simple – set up a device to act as the source, establish a connection with the test device and check for vendor-defined messages.

Testing for vendor-defined messages sounds simple, but the reality is quite different. Because the protocol specification states that any vendor-defined message must include a vendor ID, it is necessary to know or guess the expected vendor ID of the device of interest. This is important because a device would not respond to a vendor-defined message containing a correctly-guessed command but an incorrect vendor ID in the header.

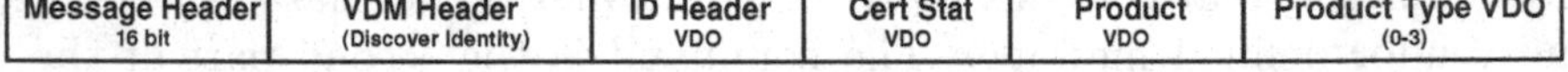

Message Header 16 bit	VDM Header (Discover Identity)	ID Header VDO	Cert Stat VDO	Product VDO	Product Type VDO (0-3)

Figure 7.   Discover Identity reply packet.

Fortunately, it is possible to leverage the Discover Identity command in the structured vendor-defined message command set shown in Table 2. This command is required by the USB Power Delivery protocol, so all devices should support the command. The command, which enables devices and cables to identify other end points, has a predefined reply packet format with a fixed number of vendor data objects and their content (Figure 7). The ID header of the 32-bit vendor data object has bits 0–15 reserved for the device USB-IF vendor ID. A connected device reveals its vendor ID upon receiving a Discover Identity command.

The protocol specification also states that structured vendor-defined messages shall only be used when an explicit contract is in place (except for a small number of cables that are not relevant in this context). The same holds true for unstructured vendor-defined messages. Thus, a device will not reply to a vendor-defined message until an explicit contract is in place (i.e., a power source configuration has been negotiated). Therefore, it is required to simulate a complete explicit contract negotiation with a test device before a vendor-defined message can be received.

This makes it necessary to simulate many messages (Figure 3) with corresponding time-outs, such as CRCReceiveTimer (maximum 1.1 ms), SenderResponseTimer (maximum 30 ms) and PSTransitionTimer (maxiumum 550 ms). Since the protocol defines the time-out values, the reply to a packet must be provided in time or the device will time out. Many of these requirements are strict, so the simulator must have a quick response, which, in turn, may render a pure software solution infeasible.

By negotiating an explicit contract with a device, it is possible to explore the existence of unstructured vendor-defined commands. Using the vendor ID captured from the response of a device to a Discover Identity command, different unstructured vendor-defined commands could be sent to the device and the responses, if any, could be examined. This can be done by brute forcing the lower 15 vendor use bits of the unstructured vendor-defined message header (Figure 5) with a fixed vendor ID for each device.

Two approaches are possible. The first is to attempt to measure the skews in the timing of device responses. The second is to test for device responses other than the expected GoodCRC message. Testing for timing skews could indicate that the device spent additional time to process a correctly-guessed unstructured vendor-defined command. However, this approach requires high resolution timers. Unfortunately, the experimental setup could only measure the time elapsed from when a packet was sent to when the response was received, which was much too inaccurate. Therefore, the second approach involving device responses other than the expected GoodCRC message was employed in the experiments.

5.   Experimental Results

Not every device with a USB Type-C connector is enabled for the USB Power Delivery protocol. If a test device with a USB Type-C connector does not respond with a GoodCRC message to the Source_Capabilities

Table 3.   Test devices with USB Type-C connectors and protocol support.

Device (Model)	Firmware Version	Protocol Revision	Exposed Vendor ID
HTC 10 (2PS6200)	1.90.401.5	2.0	0x0bb4 (HTC)
HTC U11 (2PZC100)	1.13.401.1	3.0	0x05c4 (Qualcomm)
Huawei Mate 10 Pro (BLA-L29)	8.0.0.137(C432)	2.0	0x12d1 (Huawei)
LG G5 (LG-H850)	V10i-EUR-XX MMB29M	2.0	0x0000 (Unknown)
Nokia 8 Sirocco (TA-1005)	00WW_3_10F	2.0	0x05c6 (Qualcomm)
Samsung Galaxy S9 (G960F)	G960FXXU2BRH7	3.0	0x04e8 (Samsung)

message in an explicit contract negotiation (Figure 3), then the device can be assumed to be non-protocol-enabled.

According to Section 6.2.1.1.5 of USB Power Delivery Protocol Specification Revision 3.0 (v.1.2) [8], the source shall set its highest supported specification revision in the specification revision field of the Source_Capabilities message and the sink shall reply with its highest supported specification revision in the specification revision field of the Request message (Figure 3). Because the specification states that the specification revision field value should be backwards compatible, this means the highest version can always be simulated in the first Source_Capabilities message acting as the source and the Request response from the device can then be checked.

After negotiating a complete explicit contract (Figure 3) with a test device, a Discover Identity message was sent to the device to obtain the USB-IF vendor ID from the device. Table 3 shows the test devices with USB Type-C connectors that were determined via this technique to support the USB Power Delivery protocol.

With an explicit contract in place with a test device with protocol support and its USB-IF vendor ID known, the next step was to send arbitrary protocol messages to the device and test the responses. Specifically, unstructured vendor-defined messages were sent with the vendor ID set to the appropriate value, type set to 0 (i.e., unstructured) and vendor use set to different values corresponding to commands (Figure 5).

Table 4. Huawei Mate 10 Pro (BLA-L29) message capture.

ID	Time	Role	Message	Data
284	0:41.044.922		Hard Reset	
286	0:43.577.218	Source:DFP	[0]Source_Cap	A1 11 F0 90 01 08 FE CA B7 52
290	0:43.577.879	Sink:UFP	[0]GoodCRC	41 00 BB 6C BB A8
293	0:43.580.754	Sink:UFP	[0]Request	42 10 C8 20 03 13 52 0F 95 B7
297	0:43.581.374	Source:DFP	[0]GoodCRC	A1 01 C1 AF C2 81
300	0:43.582.060	Source:DFP	[1]Accept	63 03 21 7B 00 96
303	0:43.582.586	Sink:UFP	[1]GoodCRC	41 02 97 0D B5 46
306	0:43.583.283	Source:DFP	[2]PS_RDY	A6 05 1F FD EE C9
309	0:43.583.915	Sink:UFP	[2]GoodCRC	41 04 A2 A8 D6 AF
312	0:43.737.641	Source:DFP	[0]VDM:DiscIdentity	6F 11 01 80 00 FF 76 31 6B 61
316	0:43.738.185	Sink:UFP	[0]GoodCRC	41 00 BB 6C BB A8
319	0:43.744.295	Sink:UFP	[1]VDM:DiscIdentity	4F 52 41 80 00 FF D1 12 00 EC 00 00 00 00 00 00 7E 10 01 00 00 11 80 C1 C7 56
327	0:43.745.502	Source:DFP	[1]GoodCRC	61 03 A3 19 36 A4
330	0:44.918.448	Source:DFP	[1]VDM:Unstructured	6F 13 01 00 D1 12 0D 13 06 BC
334	0:44.919.214	Sink:UFP	[1]GoodCRC	41 02 97 0D B5 46
337	0:46.507.375	Source:DFP	[2]VDM:Unstructured	6F 15 02 00 D1 12 43 49 F3 21
341	0:46.507.960	Sink:UFP	[2]GoodCRC	41 04 A2 A8 D6 AF

The responses were analyzed and any response other than the expected GoodCRC was assumed to be an attempt by the test device to reply to the random "command" it received.

A commercial USB Power Delivery protocol recorder was used to capture communications with the test devices. Table 4 shows an example capture of messages to and from the Huawei test device that was configured as the sink. The message capture shows the entire explicit contract negotiation (message IDs 286–309) and the USB-IF vendor ID discovery (message IDs 312–327), which are followed by two unstructured vendor-defined message brute force attempts (message IDs 330–334 and message IDs 337–341). Note that the Huawei device did not respond to the two unstructured vendor-defined message tests with anything other than the expected GoodCRC message.

Very few test devices responded to the brute force test. In fact, only the Samsung device replied with anything other than a GoodCRC message, and only for some messages.

Table 5 shows an example capture of messages to and from the Samsung Galaxy S9 test device that was configured as the sink. Once again, the message capture shows the entire explicit contract negotiation (message IDs 5442–5465) and the USB-IF vendor ID discovery (message IDs

Table 5.   Samsung Galaxy S9 (G960F) message capture.

ID	Time	Role	Message	Data
5440	14:36.248.230		Hard Reset	
5442	14:39.309.886	Source:DFP	[0]Source_Cap	A1 11 F0 90 01 08 FE CA B7 52
5446	14:39.310.395	Sink:UFP	[0]GoodCRC	41 00 BB 6C BB A8
5449	14:39.311.982	Sink:UFP	[0]Request	82 10 F0 C0 03 13 08 11 00 3A
5453	14:39.312.708	Source:DFP	[0]GoodCRC	A1 01 C1 AF C2 81
5456	14:39.313.284	Source:DFP	[1]Accept	63 03 21 7B 00 96
5459	14:39.313.979	Sink:UFP	[1]GoodCRC	41 02 97 0D B5 46
5462	14:39.314.462	Source:DFP	[2]PS_RDY	A6 05 1F FD EE C9
5465	14:30.315.049	Sink:UFP	[2]GoodCRC	41 04 A2 A8 D6 AF
5468	14:39.471.248	Source:DFP	[0]VDM:DiscIdentity	6F 11 01 80 00 FF 7C 31 6B 61
5472	14:39.471.866	Sink:UFP	[0]GoodCRC	41 00 BB 6C BB A8
5475	14:39.476.288	Sink:UFP	[1]VDM:DiscIdentity	8F 42 41 80 00 FF E8 04 00 D1 00 00 00 00 00 00 60 68 C2 B2 A2 9E
5482	14:39.477.131	Source:DFP	[1]GoodCRC	61 03 A3 19 36 A4
5485	14:40.650.372	Source:DFP	[1]VDM:Unstructured	6F 13 01 00 E8 04 E6 2B 56 46
5489	14:40.651.199	Sink:UFP	[1]GoodCRC	41 02 97 0D B5 46
5492	14:40.654.796	Sink:UFP	[2]VDM:Unstructured	4F 14 41 00 E8 04 FD AA CE 68
5496	14:40.655.473	Source:DFP	[2]GoodCRC	61 05 96 BC 55 4D
5499	14:41.828.228	Source:DFP	[2]VDM:Unstructured	6F 15 02 00 E8 04 A8 71 A3 DB
5503	14:41.829.056	Sink:UFP	[2]GoodCRC	41 04 A2 A8 D6 AF
5506	14:41.833.325	Sink:UFP	[3]VDM:Unstructured	4F 56 42 00 E8 04 00 00 00 00 00 00 00 00 00 00 00 00 00 00 00 00 00 34 A1 0A 25
5514	14:41.834.581	Source:DFP	[3]GoodCRC	61 07 BA DD 5B A3
5517	14:43.008.455	Source:DFP	[3]VDM:Unstructured	6F 17 02 00 E8 04 C8 22 63 A1
5521	14:43.009.071	Sink:UFP	[3]GoodCRC	41 06 8E C9 D8 41
5524	14:43.013.435	Sink:UFP	[4]VDM:Unstructured	4F 58 42 00 E8 04 00 00 00 00 00 00 00 00 00 00 00 00 00 00 00 00 00 84 AD C5 F6
5532	14:43.014.693	Source:DFP	[4]GoodCRC	61 09 BD F0 E3 44
5535	14:44.180.619	Source:DFP	[4]VDM:Unstructured	6F 19 03 00 E8 04 CC FB EF A6
5539	14:44.181.134	Sink:UFP	[4]GoodCRC	41 08 89 E4 60 A6
5542	14:45.761.683	Source:DFP	[5]VDM:Unstructured	6F 1B 02 00 E8 04 C9 CF 93 64
5546	14:45.762.289	Sink:UFP	[5]GoodCRC	41 0A A5 85 6E 48
5549	14:45.766.649	Sink:UFP	[5]VDM:Unstructured	4F 5A 42 00 E8 04 0D DA 95 63 4A 97 17 B5 F5 34 11 47 53 7E C9 E9 8C 35 3F 0E
5557	14:45.767.917	Source:DFP	[5]GoodCRC	61 0B 91 91 ED AA
5560	14:46.933.424	Source:DFP	[6]VDM:Unstructured	6F 1D 01 00 E8 04 87 95 66 F9
5564	14:46.934.042	Sink:UFP	[6]GoodCRC	41 0C 90 20 0D A1
5567	14:46.937.851	Sink:UFP	[6]VDM:Unstructured	4F 1C 41 00 E8 04 3C E1 BE 58
5571	14:46.938.566	Source:DFP	[6]GoodCRC	61 0D A4 34 8E 43
5574	14:48.114.825	Source:DFP	[7]VDM:Unstructured	6F 1F 02 00 E8 04 09 69 13 91
5578	14:48.115.442	Sink:UFP	[7]GoodCRC	41 0E BC 41 03 4F
5581	14:48.119.820	Sink:UFP	[7]VDM:Unstructured	4F 5E 42 00 E8 04 0D DA 95 63 4A 97 17 B5 F5 34 11 47 53 7E C9 E9 37 31 C6 1C
5589	14:48.121.075	Source:DFP	[7]GoodCRC	61 0F 88 55 80 AD
5592	14:49.303.445	Source:DFP	[0]VDM:Unstructured	6F 11 03 00 E8 04 0D B0 9F 96
5596	14:49.304.274	Sink:UFP	[0]GoodCRC	41 00 BB 6C BB A8
5599	14:50.881.168	Source:DFP	[1]VDM:Unstructured	6F 13 02 00 E8 04 08 84 E3 54
5603	14:50.881.789	Sink:UFP	[1]GoodCRC	41 02 97 0D B5 46
5606	14:50.886.156	Sink:UFP	[0]VDM:Unstructured	4F 50 42 00 E8 04 60 B3 A9 5A 65 3F 48 3C 3A D6 13 DC 2D 32 8D 16 F6 75 A3 FE
5614	14:50.887.366	Source:DFP	[0]GoodCRC	61 01 8F 78 38 4A

5468–5482). These are followed by the first unstructured vendor-defined message test (message ID 5485). The sent message has an unstructured vendor-defined message header of 0x04e80001, which is decoded according to Figure 5 as vendor ID: 0x04e8, type: 0 and vendor use: 0x0001 (15-bit value).

Note that this unstructured vendor-defined message received a response other that the GoodCRC (message ID 5492). The response has an unstructured vendor-defined message header of 0x04e80041, which is decoded according to Figure 5 as vendor ID: 0x04e8, type: 0 and vendor use: 0x0041. This message appears to be a reply with no additional data (i.e., vendor data objects).

A similar situation is seen for message 5499 with vendor use: 0x0002, whose response (message ID 5506) has vendor use: 0x0042 and four additional vendor data objects: 0x00000000 0x00000000 0x00000000 and 0x00000000.

The two vendor use command/reply pairs of 0x0001/0x0041 and 0x0002/0x0042 imply that bit 6 (0x0040) may be an ACK bit. If the unstructured headers are interpreted as structured headers (Figure 6), then bits 6–7 correspond to type where 0x1 (bit 6 set) corresponds to an ACK. Of course, the real situation is not clear, but it does appear that the vendor may have mixed the two types of vendor-defined message headers.

Investigating further, the response (message ID 5506) with vendor use set to 0x0042 also has four additional vendor data objects: 0x00000000 0x00000000 0x00000000 and 0x00000000. This appears to be data sent back to the source side from the sink. All the vendor data objects contain zeroes in the replies to two consecutive messages with vendor use set to 0x0002 (message IDs 5499 and 5517).

However, when a different message (message ID 5535) is sent to the device with vendor use set to 0x0003, then a completely different reply is received with vendor use set to 0x0002 (message ID 5542) and four vendor data objects: 0x6395da0d 0xb517974a 0x471134f5 and 0xe9c97e53 (message ID 5549). Sending message 5535 again (message ID 5574) yields the same four vendor data objects (message ID 5581). However, another message with vendor use set to 0x0003 (message ID 5592) once again changes the vendor data objects for vendor use set to 0x0002. Specifically, the four vendor data objects are: 0x5aa9b360 0x3c483f65 0xdc13d63a and 0x168d322d (message ID 5606).

It appears that data in the form of vendor data objects is received from the device and different data is received when sending a specific message with vendor use set to 0x0003. The four vendor data objects appear to change in pseudorandom order. Another observation is that,

Table 6.   Samsung Galaxy S9 (G960F) message capture.

ID	Time	Role	Message	Data
162	0:06.589.154	Source:DFP	[1]VDM:Unstructured	6F 13 01 00 E8 04 E6 2B 56 46
166	0:06.589.982	Sink:UFP	[1]GoodCRC	41 02 97 0D B5 46
169	0:06.594.059	Sink:UFP	[1]VDM:Unstructured	4F 12 41 00 E8 04 5D 5F 8E E7
173	0:06.594.675	Source:DFP	[1]GoodCRC	61 03 A3 19 36 A4
176	0:06.629.222	Source:DFP	[2]VDM:Unstructured	6F 55 02 00 E8 04 1C 47 B3 AB 2E F3 7B AE
				F9 09 79 82 02 3B C6 BB 1A D4 E8 41
184	0.06.630.376	Sink:UFP	[2]GoodCRC	41 04 A2 A8 D6 AF
187	0:06.635.264	Sink:UFP	[2]VDM:Unstructured	4F 54 42 00 E8 04 1C 47 B3 AB 2E F3 7B AE
				F9 09 79 82 02 3B C6 BB 51 65 55 63
195	0:06.636.524	Source:DFP	[2]GoodCRC	61 05 96 BC 55 4D

when a message is sent with vendor use set to 0x0002 along with four random vendor data objects (0xabb3471c, 0xae7bf32e, 0x827909f9, 0xbbc63b02), a reply is received with the same vendor data objects (Table 6). This implies that a message with vendor use set to 0x0002 corresponds to an initialization command. Repeating the messages with vendor use set to 0x0003 and 0x0002 gives different vendor data objects, which may correspond to some form of encryption or obfuscation.

Sending two identical runs of the messages in Table 5 gives the same results and any randomization of the four vendor data objects sent with vendor use set to 0x0002 yields seemingly random reply vendor data objects when intermingled with messages with vendor use set to 0x0003. This strengthens the belief that encryption is in place and that the message with vendor use set to 0x0002 is either transmitting a key or an initialization vector for a symmetric cipher.

Because the results indicate that Samsung devices respond to vendor-defined messages in the USB Power Delivery protocol, additional experiments were conducted to confirm the results. The experiments employed a special factory test device called the Samsung Anyway S103 (Figure 8). This device enables a console interface provided by the device bootloader, which is useful for debug logging and other activities. The same console can be reached via a custom USB connector and a simple RS232-to-USB serial converter on older devices with micro-USB connectors [3]. Alendal et al. [1] employed this type of connection to demonstrate an exploit targeting Samsung devices with a certain security vulnerability. The exploit assisted in bypassing a security feature in the devices. This demonstrates the importance of expanding the attack surface of a device by enabling the factory test feature.

Figure 8. Samsung Anyway S103.

The special factory device was hard to obtain because it is usually provided to Samsung device repair shops and similar outlets. However, a factory device was procured to communicate with the Samsung test device using the USB Power Delivery protocol. Table 7 shows a message capture with the Samsung Anyway S103 and Samsung Galaxy S9 configured as the source and sink, respectively (the vendor data objects are partially redacted). Note that the communications in the message capture did not involve an explicit contract negotiation as required in the protocol specification. Instead, immediate vendor-defined message communications were conducted using the discovered vendor-defined messages. The capture corresponds to a vendor-defined message with vendor use set to 0x0001, followed by a vendor-defined message with vendor use set to 0x0002 that provides four pseudorandom vendor data objects. These are followed by several vendor-defined messages with vendor use set to 0x0003, each containing four vendor data objects with seemingly pseudorandom data.

Next, the Samsung Anyway S103 factory device was removed as the source and a blind replay from the source side of the communications was attempted. The idea was that, if the source messages from the Samsung Anyway S103 device were replayed and the same sink messages were received from the test device, then the Samsung Anyway S103 device was essentially being emulated. This test was an immediate success.

Table 7.   Samsung Anyway S103 and Samsung Galaxy S9 message capture.

ID	Time	Role	Message	Data
1	0:03.900.730	Source:DFP	[0]VDM:DiscIdentity	6F 11 01 80 00 FF 76 31 6B 61
5	0:03.901.546	Sink:UFP	[0]GoodCRC	41 00 BB 6C BB A8
8	0:03.905.272	Sink:UFP	[0]VDM:DiscIdentity	8F 40 41 80 00 FF E8 04 00 D1 00 00 00 00 00 00 60 68 05 22 9E 4A
15	0:03.906.336	Source:DFP	[0]GoodCRC	61 01 8F 78 38 4A
18	0:03.906.881	Source:DFP	[1]VDM:Unstructured	6F 13 01 00 E8 04 E6 2B 56 46
22	0:03.907.590	Sink:UFP	[1]GoodCRC	41 02 97 0D B5 46
25	0:03.912.440	Sink:UFP	[1]VDM:Unstructured	4F 12 41 00 E8 04 5D 5F 8E E7
29	0:03.913.109	Source:DFP	[1]GoodCRC	61 03 A3 19 36 A4
32	0:03.913.649	Source:DFP	[2]VDM:Unstructured	6F 55 02 00 E8 04 0C DD BB FF REDACTED
40	0:03.914.888	Sink:UFP	[2]GoodCRC	41 04 A2 A8 D6 AF
43	0:03.919.998	Sink:UFP	[2]VDM:Unstructured	4F 54 42 00 E8 04 0C DD BB FF REDACTED
51	0:03.921.093	Source:DFP	[2]GoodCRC	61 05 96 BC 55 4D
54	0:03.922.149	Source:DFP	[3]VDM:Unstructured	6F 57 03 00 E8 04 E6 A9 7F 72 94 CE B1 B6 54 BA B7 75 6A F1 89 B8 01 65 20 E8
62	0:03.923.388	Sink:UFP	[3]GoodCRC	41 06 8E C9 D8 41
65	0:03.931.556	Sink:UFP	[3]VDM:Unstructured	4F 56 43 00 E8 04 9F B2 F5 F9 F1 68 E2 AF E5 AA 22 73 D0 77 6A 2E B6 3A A9 FB
73	0:03.932.759	Source:DFP	[3]GoodCRC	61 07 BA DD 5B A3
76	0:03.934.596	Source:DFP	[4]VDM:Unstructured	6F 59 03 00 E8 04 F7 96 A6 2A 08 BB A9 6E 38 40 E4 AF 33 43 7A 23 E6 D7 A8 E9
84	0:03.935.837	Sink:UFP	[4]GoodCRC	41 08 89 E4 60 A6
87	0:03.942.701	Sink:UFP	[4]VDM:Unstructured	4F 58 43 00 E8 04 9A 01 DB AE 9A 39 26 77 B0 A8 2D 11 A2 C1 76 80 1E 08 1E C2
95	0:03.943.902	Source:DFP	[4]GoodCRC	61 09 BD F0 E3 44

The key result is that the same console reached on micro-USB Samsung devices was enabled without the assistance of the Samsung Anyway S103 factory device.

The successful message replay strengthens the belief that encryption is involved and that the first four vendor data objects in the vendor-defined message with vendor use set to 0x0002 are crucial to initialization. These vendor data objects could correspond to an initialization vector or perhaps even the key to a symmetric cipher. However, experiments with several symmetric ciphers using the four vendor data objects as the key to decrypt vendor data objects in messages with the vendor use set to 0x0003 did not yield positive results.

6. Conclusions

The principal contribution of this research is a testing methodology and implementation for revealing and analyzing proprietary USB Power Delivery protocol messages. The experimental results demonstrate that at least one common mobile device, the Samsung Galaxy S9, is amenable to the testing methodology. In particular, the device responds to certain vendor-defined messages and the responses indicate the use of encryption, which raises the possibility of capturing initialization vectors and keys for symmetric ciphers. Another important result is the ability to enable factory device features in a test device in order to obtain valuable log data from the device and to widen its attack surface.

Future research will continue the investigation of vendor-defined messages in the USB Power Delivery protocol. Since vendors may also implement hidden features in other parts of the protocol, a promising approach is to investigate the role of the sink device that consumes power. Connecting two devices that typically serve as sinks – like two mobile phones – causes one device to assume the source role and provide power to the other device. This source-sink relationship could be exploited to expand the attack surface or even to directly acquire data.

Future research will also investigate potential security vulnerabilities. This is challenging because it is not known how to instrument a USB Power Delivery chip for feedback (e.g., if it crashes or demonstrates anomalous behavior). An alternative approach is to conduct a source code review or extract the chip firmware and apply reverse engineering techniques. Another approach is to analyze device-side communications with the USB Power Delivery chip, which could reveal interesting features or vulnerabilities in the chip logic as well in the operating system.

The popularity of USB Type-C connectors is increasing and large numbers of consumer devices will support the USB Power Delivery protocol. It is hoped that this work will stimulate research on the protocol and its implementations to advance device security and forensics.

Acknowledgement

This research was sponsored by the Norwegian Research Council IK-TPLUSS Program under the Ars Forensica Project No. 248094/O70.

References

[1] G. Alendal, G. Dyrkolbotn and S. Axelsson, Forensic acquisition – Analysis and circumvention of Samsung secure boot enforced common criteria mode, *Digital Investigation*, vol. 24(S), pp. S60–S67,

2018.

[2] G. Alendal, C. Kison and modg, Got HW Crypto? On the (In)Security of a Self-Encrypting Drive Series, Cryptology ePrint Archive, Report 2015/1002 (eprint.iacr.org/2015/1002), 2015.

[3] N. Artenstein, Exploiting Android S-Boot: Getting arbitrary code exec in the Samsung bootloader (1/2), *Information Security Newspaper*, March 3, 2017.

[4] Chindi.ap (commons.wikimedia.org/wiki/User:Chindi.ap), 2019.

[5] H. Reydarns, V. Lauwereys, D. Haeseldonckx, P. van Willigenburg, J. Woudstra and S. De Jonge, The development of a proof of concept for a smart DC/DC power plug based on USB Power Delivery, *Proceedings of the Twenty-Second Conference on the Domestic Use of Energy*, 2014.

[6] T10 Technical Committee of the International Committee on Information Technology Standards, SCSI Operation Codes (www.t10.org/lists/op-num.htm), 2015.

[7] USB Implementers Forum, Getting a Vendor ID, Beaverton, Oregon (www.usb.org/getting-vendor-id), 2019.

[8] USB Implementers Forum, USB Power Delivery, Beaverton, Oregon (www.usb.org/document-library/usb-power-delivery), 2019.

Chapter 7

DETECTING ANOMALIES IN PROGRAMMABLE LOGIC CONTROLLERS USING UNSUPERVISED MACHINE LEARNING

Chun-Fai Chan, Kam-Pui Chow, Cesar Mak and Raymond Chan

Abstract   Supervisory control and data acquisition systems have been employed for decades to communicate with and coordinate industrial processes. These systems incorporate numerous programmable logic controllers that manage the operations of industrial equipment based on sensor information. Due to the important roles that programmable logic controllers play in industrial facilities, these microprocessor-based systems are exposed to serious cyber threats.

This chapter describes an innovative methodology that leverages unsupervised machine learning to monitor the states of programmable logic controllers to uncover latent defects and anomalies. The methodology, which employs a one-class support vector machine, is able to detect anomalies without being bound to specific scenarios or requiring detailed knowledge about the control logic. A case study involving a traffic light simulation demonstrates that anomalies are detected with high accuracy, enabling the prompt mitigation of the underlying problems.

Keywords: Programmable logic controllers, anomaly detection, machine learning

1.   Introduction

Supervisory control and data acquisition (SCADA) systems have been employed for decades to manage and control critical infrastructure assets. With human lives and the economy at stake, SCADA system failures – whether due to accidents or attacks – cannot be tolerated. Therefore, it is vital to detect SCADA system anomalies and implement effective mitigation strategies.

© IFIP International Federation for Information Processing 2019
Published by Springer Nature Switzerland AG 2019
G. Peterson and S. Shenoi (Eds.): Advances in Digital Forensics XV, IFIP AICT 569, pp. 119–130, 2019.
https://doi.org/10.1007/978-3-030-28752-8_7

Programmable logic controllers (PLCs) are the workhorses of SCADA systems. These microprocessor-based systems implement programmable logic that processes input signals from sensors that measure system/environment state to produce output signals that are transmitted to actuators as well as other programmable logic controllers that operate and manage industrial equipment and processes. Programmable logic controllers are typically small, rugged, specialized devices designed to perform specific control tasks, often operating in harsh environments with extreme temperatures and strong vibrations. Industrial systems may have tens to hundreds of programmable logic controllers. Large infrastructure assets such as power grids and oil and gas pipelines have thousands of programmable logic controllers.

Programmable logic controllers are exposed to inadvertent and malicious threats that can impact their ability to safely operate industrial systems and facilities. The most common inadvertent threats are posed by control program implementation bugs. Malicious threats include memory read/write logic attacks [20, 21], malware worms [5, 6, 16], time bombs [1, 7], and stop and start attacks [22]. These threats make it imperative to develop security solutions for monitoring the states of programmable logic controllers to uncover latent defects and anomalies.

Unfortunately, the limited computational and storage resources of programmable logic controllers make it difficult to deploy conventional security measures such as firewalls and intrusion detection systems. Novel and efficient methodologies are required to detect anomalous controller behavior in real time, and help support prompt mitigations and forensic investigations of incidents [9, 22].

Machine learning, which has been employed with much success in intrusion and anomaly detection systems for traditional computing and networking infrastructures, is a promising approach for developing similar systems for programmable logic controllers. Supervised learning, which takes in training data with labeled outcomes, is oriented towards data clustering and classification. Unsupervised learning, which takes in unlabeled data, is geared towards outlier detection. In both cases, a mathematical model is generated from the training data and the model serves as a classifier for new data. Either model can be used for anomaly detection.

It is difficult to apply supervised learning to detect attacks on programmable logic controllers due to the lack of genuine attack data; additionally, the problem spaces (numbers of attack patterns) are large and simulating every attack pattern to generate data is infeasible. In contrast, unsupervised learning uses datasets without labels [12, 14]. A training dataset covering normal behavior is created and normalized to construct a model that identifies outliers.

Anomaly detection is conceptually identical to outlier detection, which makes unsupervised learning ideal for the problem at hand. In fact, outlier identification is virtually equivalent to applying unary classification with respect to good cases.

A one-class support vector machine is a special case of a support vector machine with unary classification [17]. In this approach, data points are grouped using correlations that are computed to yield the normal state class. The region corresponding to normal state class data is used to assess if a new data point is an outlier. This approach is essentially a sophisticated regression test where the training data is processed collectively. It is especially appropriate when the training dataset mainly comprises normal state data and very little anomalous data. Indeed, the approach is well-suited to anomaly detection in programmable logic controllers because attacks are rare and attack data is hard to come by whereas normal data is readily captured during day-to-day operations.

This chapter describes a methodology that leverages unsupervised machine learning to monitor the states of programmable logic controllers to uncover latent defects and anomalies. The methodology, which employs a one-class support vector machine, can detect anomalies without being bound to specific scenarios or requiring detailed knowledge about the control logic. In addition to conventional data capture methods, the methodology leverages an additional security block in a programmable logic controller to detect anomalies [1]. The historian is also employed to store timestamped programmable logic controller state information (i.e., key memory address values) for anomaly/attack analyses and forensic investigations. A traffic light simulation case study employing a Siemens S7-1212C programmable logic controller demonstrates that anomalies are detected with high accuracy.

2. Related Work

Garitano et al. [2] have reviewed several anomaly detection methodologies and conclude that network intrusion detection systems may not be able to efficiently detect attacks on industrial control systems. Furthermore, since programmable logic controllers typically have limited computational resources, implementing host-based intrusion detection systems is generally infeasible.

Hsu et al. [4] have evaluated several machine learning algorithms on datasets comprising normal operational data from SCADA networks.

Their results demonstrate that machine learning algorithms are able to accurately detect most attacks.

Schuster et al. [10] conducted anomaly detection experiments in two plant process control networks using one-class support vector machines and isolation forest classifiers. Their studies revealed that network traffic data is inadequate for training purposes when sufficient programmable logic controller traffic is not available.

Wu and Nurse [18] have observed that valuable information can be obtained by monitoring the memory addresses of programmable logic controllers, regardless of whether the controllers were executing normally or were under attack. They also evaluated the use of a programmable logic controller logger as a forensic tool that continuously polls the memory variables in a running programmable logic controller.

Yau and Chow [20, 21] have proposed two approaches for detecting attacks on programmable logic controllers. One approach applies machine learning to logged data of pre-selected memory values of a programmable logic controller to detect abnormal operations [20]. The other approach employs a control program logic change detector that leverages anomaly detection rules to detect and record undesirable events [19].

Both the approaches require knowledge about the control logic before monitoring procedures can be applied. They may, therefore, be impractical because the personnel responsible for monitoring the security of SCADA systems are typically not involved in SCADA system development. In addition, remote monitoring of programmable logic controllers via active polling imposes network overhead that is unacceptable in industrial control system environments.

To overcome these challenges, Chan et al. [1] proposed the incorporation of a security block module to support programmable logic controller logging and attack detection capabilities. Specifically, they installed a security block (i.e., programmable logic controller code) on the device to capture selected memory content and other internal device information for monitoring purposes. This approach can help detect programmable logic controller memory read-write logic attacks with high accuracy while maintaining a low network footprint. In addition, the security block can verify the number of data blocks installed in a programmable logic controller to detect worm attacks, which is more efficient than traditional network memory address value polling method using libnodave [3] or the Siemens Step 7 library [8].

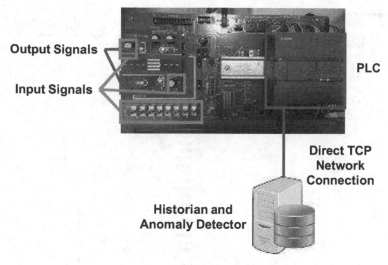

Figure 1. Experimental setup.

3. Anomaly Detection Case Study

This section describes the experimental setup and the methodology for detecting anomalous programmable logic controller operations.

3.1 Experimental Setup

Figure 1 shows the experimental setup. A Siemens S7-1212C programmable logic controller was installed with a traffic light control program that manages interactions between switches and the sequencing and durations of traffic lights. In addition to the standard traffic control light program, the programmable logic controller was equipped with a security block that transmitted input, output and memory address values to a historian via a direct TCP connection. All this information was recorded in a log file by the historian for anomaly detection and forensic analysis.

Rogue attacks on the programmable logic controller were executed by incorporating attack logic in the device. Upon receiving certain input signals, the program logic altered output signals to launch the attacks.

The objective of the experiment was to detect anomalous behavior. Events such as direct attacks, hardware failures and implementation bugs produce anomalies. By attaching timestamps to the events, anomalous situations can also be investigated retroactively by examining the data maintained by the historian.

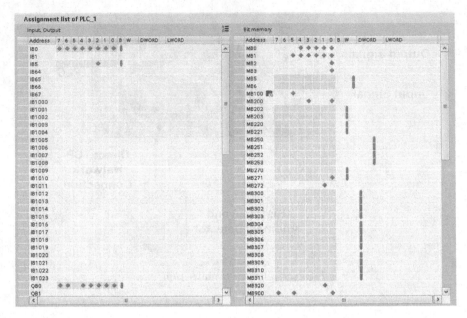

Figure 2.   Assignment list for the traffic light system with a security block.

3.2   Anomaly Detection Methodology

In order to detect anomalies, it is necessary to capture adequate amounts of useful data. Since there is no prior information about the programmable logic controller logic, it is necessary to determine which memory addresses are referenced by the controller logic.

If the source code of the traffic light program is available, the code can be loaded into TIA [15], an integrated development environment for Siemens programmable logic controllers, which creates an assignment list that contains all the referenced memory addresses. Figure 2 shows the assignment list for the traffic light program. Because the security block was configured to use memory block addresses MB200 to MB900, these addresses were deemed to be irrelevant and were, therefore, ignored in the data capture.

If the source code is not available, then it is necessary to capture the contents of memory addresses during normal operation. The memory capture process is repeated for small memory blocks until the contents of all the memory addresses have been captured. Next, the memory addresses whose contents do not change are eliminated based on the assumption that their inactivity implies that they have no impact on programmable logic controller behavior. The assumption is reasonable for programs that do not flip and restore memory addresses during a cycle, and have no external dependencies. This turned out to be the case for the traffic light simulation program.

Since the source code was available in the experiment, the Siemens TIA integrated development environment was used to identify the memory addresses of interest in the programmable logic controller.

After the memory addresses have been identified, several approaches can be used to capture information about programmable logic controller status. One approach is to use a network sniffer or mirror port in a network device to capture network traffic to and from the programmable logic controller. Another approach is to actively poll memory address values using an external program [19]. Yet another approach is to use a security block to transmit internal programmable logic controller data [1] to a historian.

The approach adopted in this work was to capture and analyze input and output signals and memory values using a security block. One reason is that, in many real-world deployments (as in the case of the traffic control experiment), a programmable logic controller has minimal external network traffic – because it is directly connected to input/output ports, not all signals generate network traffic traces during normal operation. In addition, stateful information about programmable logic controller operations may not be transferred to an external device such as a historian for storage. Thus, network traffic captures alone would not provide adequate information about the programmable logic controller.

Since different combinations of memory address values may represent different program states, it is important to ensure that the captured values are consistent within a programmable logic controller execution cycle. However, using an external program (e.g., Snap7 [8]) over a network to query memory address values does not ensure their consistency due to network latency and programmable logic controller operating system delays. In addition, continuously polling multiple memory addresses imposes overhead on a programmable logic controller that may degrade its performance.

These challenges are overcome using a security block to produce a consistent snapshot of memory values in every cycle. Other advantages of the security block over active polling are higher levels of correlation between memory addresses, less lag and missing state data, and accurate timestamp information for forensic analyses.

Since the data transferred from a security block is in the form of a tokenized byte stream, the byte stream has to be converted back to its original data types (e.g., integer and boolean) for input to a machine learning model.

The popular OCSVM outlier detection machine learning model [10] was employed to detect anomalies. The specific OCSVM model used was from Scikit-learn [13], an open-source software package that provides several machine learning libraries written in Python. The formatted input data was provided to the OCSVM model libraries.

The OCSVM parameters were optimized to increase model accuracy before it underwent training and testing. This was achieved by applying a portion of the original dataset to conduct an iterative search for the best parameters.

The following optimized parameter settings were employed:

- **Kernel:** This parameter specifies the non-linear function used by the support vector machine to project the hyperspace to a higher dimension. The optimal rbf kernel setting was used.

- **Degree:** This parameter specifies the degree of the polynomial kernel function. The optimal degree setting of four was used.

- **Coef0:** This parameter is not significant for the rbf kernel. The optimal setting of zero was used.

- **Nu:** This parameter specifies the maximum number of training examples that can be misclassified and the minimum fraction of training examples for the support vector. The optimal setting of 0.001 was used.

The next step involved the generation of anomalous events and data collection. Programs were executed to inject attack traffic into the programmable logic controller to simulate real attacks. Normal and attack data collected by the historian were used as samples.

The following metrics for benchmarking the accuracy of machine learning techniques [11] were used to assess the effectiveness of anomaly detection:

- **Precision:** Precision is defined as the ratio of true positives to the sum of true positives and false positives $(= \frac{TP}{TP+FP})$. It measures the ability of a classifier to not misclassify negative samples as positive samples. The precision ranges from one (best) to zero (worst).

- **Recall:** Recall is defined as the ratio of true positives to the sum of true positives and false negatives $(= \frac{TP}{TP+FN})$. It measures the ability of a classifier to identify all the positive samples. The recall ranges from one (best) to zero (worst).

Table 1. Classification results for various performance metrics.

	Records	Precision	Recall	F1 Score
Training Set	922,162	1.00	0.94	0.97
Testing Set 1	505,041	0.97	0.93	0.94
Testing Set 2	588,573	0.95	0.94	0.94
Testing Set 3	471,207	0.97	0.95	0.96

- **F1 Score:** The F1 score is a weighted average of precision and recall $(= \frac{TP+FN}{TP+FP})$. The higher the score, the better the ability of a classifier to detect negative samples while maintaining a low false positive rate. The F1 score ranges from one (best) to zero (worst).

Table 1 shows the anomaly detection results for the training set and three testing sets. Good results were obtained. The precision for the three testing sets ranges from 0.95 to 0.97; the recall ranges from 0.93 to 0.95; and the F1 score ranges from 0.94 to 0.96.

4. Discussion

In industrial control environments, it is difficult to obtain attack data and little, if any, details are available about the internal logic of programmable logic controllers. Since adequate amounts of normal operational data are available, the solution to detecting anomalies caused by attacks is to employ a machine learning technique create a model of normal behavior and use the trained model to identify anomalous behavior. The experimental results demonstrate that the trained detector was able to recognize normal behavior with a low error rate. Thus, it would be effective as a monitoring mechanism for detecting unknown attacks and unanticipated failures.

The experiments assumed that the training dataset contained only normal scenarios, without any anomalous events. This requires the number of normal scenarios in the dataset to be substantial enough to be distinguishable from anomalous outliers. The experiments revealed that insufficient amounts of training data about normal scenarios yield high false positive rates. Therefore, a large normal dataset must be used during the training phase.

Another observation is that the model parameters have large impacts on the accuracy of detection. A previous study with the simulated traffic light system [21] revealed that the default parameter settings yield modest results. In contrast, the experiments described in this chapter demonstrate that good results are obtained by using a small dataset in an iterative search for optimal model parameters and then applying the model with the optimized parameters to larger datasets for training and testing.

The logging mechanism implemented by the historian maintains precise timestamps of programmable logic controller memory status. The timestamped information coupled with the trained anomaly detection model can significantly advance forensic investigations. For example, when the anomaly detection model triggers an alert with a concrete timestamp, a forensic investigator can narrow down the time and duration of the incident, and look up and recreate the programmable logic controller memory status and behavior using the data stored by the historian.

However, the proposed approach has some limitations. First, the analysis of memory addresses is not scalable. If large numbers of memory addresses are used by a programmable logic controller, then a filtering mechanism would be required to reduce the number of features considered by the machine learning model. Second, if a programmable logic controller stores its state data on an external device during its execution cycle, this data must be obtained and verified to ensure accurate detection. Finally, unsupervised learning requires a large and rich normal dataset for model training.

5.   Conclusions

The lack of genuine attack data and the difficulty in generating simulated attack data render unsupervised learning well-suited to developing anomaly detection systems for programmable logic controllers. The proposed anomaly detection methodology, which employs a one-class support vector machine, accurately detects anomalies without being bound to specific scenarios or requiring detailed knowledge about the control logic. The methodology leverages an additional security block in a programmable logic controller to detect anomalies and employs the historian to store timestamped programmable logic controller state information (i.e., key memory address values) to support anomaly/attack analyses and forensic investigations. Experimental results with a traffic light simulation system employing a Siemens S7-1212C programmable logic controller demonstrate that anomalies are detected with high accuracy.

Future research will focus on implementing increased state awareness based on live programmable logic controller memory analysis to enhance anomaly detection. Efforts will also concentrate on tuning the unsupervised learning methodology to enhance performance metrics such as precision, recall and the F1 score.

References

[1] C. Chan, K. Chow, S. Yiu and K. Yau, Enhancing the security and forensic capabilities of programmable logic controllers, in *Advances in Digital Forensics XIV*, G. Peterson and S. Shenoi (Eds.), Springer, Cham, Switzerland, pp. 351-367, 2018.

[2] I. Garitano, R. Uribeetxeberria and U. Zurutuza, A review of SCADA anomaly detection systems, *Proceedings of the Sixth International Conference on Soft Computing Models in Industrial and Environmental Applications*, pp. 357–366, 2011.

[3] T. Hergenhahn, libnodave (`sourceforge.net/projects/libnoda ve`), 2014.

[4] J. Hsu, D. Mudd and Z. Thornton, Project Report – SCADA Anomaly Detection, Department of Electrical and Computer Engineering, Mississippi State University, Mississippi State, Mississippi (`www.ece.uah.edu/~thm0009/icsdatasets/MSU_SCADA_ Final_Report.pdf`), 2014.

[5] S. Karnouskos, Stuxnet worm impact on industrial cyber-physical system security, *Proceedings of the Thirty-Seventh Annual Conference of the IEEE Industrial Electronics Society*, pp. 4490–4494, 2011.

[6] J. Klick, S. Lau, D. Marzin, J. Malchow and V. Roth, Internet-facing PLCs as a network backdoor, *Proceedings of the IEEE Conference on Communications and Network Security*, pp. 524–532, 2015.

[7] Langner, A time bomb with fourteen bytes, Dover, Delaware (`www. langner.com/2011/07/a-time-bomb-with-fourteen-bytes`), July 21, 2011.

[8] D. Nardella, Step 7 Open Source Ethernet Communications Suite, Bari, Italy (`snap7.sourceforge.net`), 2016.

[9] S. Nazir, S. Patel and D. Patel, Assessing and augmenting SCADA cyber security: A survey of techniques, *Computers and Security*, vol. 70, pp. 436-454, 2017.

[10] F. Schuster, F. Kopp, A. Paul and H. Konig, Attack and fault detection in process control communications using unsupervised machine learning, *Proceedings of the Sixteenth International Conference on Industrial Informatics*, pp. 433–438, 2018.

[11] Scikit-learn Project, scikitlearn.metrics: Metrics (`scikit-learn.org/stable/modules/classes.html#sklearn-metrics-metrics`), 2016.

[12] Scikit-learn Project, Novelty and Outlier Detection (`scikit-learn.org/stable/modules/outlier_detection.html#outlier-detection`), 2017.

[13] Scikit-learn Project, sklearn.svm.OneClassSVM (`scikit-learn.org/stable/modules/generated/sklearn.svm.OneClassSVM.html`), 2017.

[14] Scikit-learn Project, An Introduction to Machine Learning with scikitlearn (`scikit-learn.org/stable/tutorial/basic/tutorial.html`), 2018.

[15] Siemens, Totally Integrated Automation Portal, Nuremberg, Germany, 2019.

[16] R. Spenneberg, M. Bruggemann and H. Schwartke, PLC-blaster: A worm living solely in the PLC, presented at *Black Hat USA*, 2016.

[17] Wikipedia, One-Class Classification (`en.wikipedia.org/wiki/One-class_classification`), 2018.

[18] T. Wu and J. Nurse, Exploring the use of PLC debugging tools for digital forensic investigations of SCADA systems, *Journal of Digital Forensics, Security and Law*, vol. 10(4), pp. 79–96, 2015.

[19] K. Yau and K. Chow, PLC forensics based on control program logic change detection, *Journal of Digital Forensics, Security and Law*, vol. 10(4), pp. 59–68, 2015.

[20] K. Yau and K. Chow, Detecting anomalous programmable logic controller events using machine learning, in *Advances in Digital Forensics XIII*, G. Peterson and S. Shenoi (Eds.), Springer, Cham, Switzerland, pp. 81–94, 2017.

[21] K. Yau, K. Chow, S. Yiu and C. Chan, Detecting anomalous behavior of PLCs using semi-supervised machine learning, *Proceedings of the IEEE Conference on Communications and Network Security*, pp. 580–585, 2017.

[22] E. Yilmaz and S. Gonen, Attack detection/prevention system against cyber attacks on industrial control systems, *Computers and Security*, vol. 77, pp. 94–105, 2018.

III

FILESYSTEM FORENSICS

Chapter 8

CREATING A MAP OF USER DATA IN NTFS TO IMPROVE FILE CARVING

Martin Karresand, Asalena Warnqvist, David Lindahl, Stefan Axelsson and Geir Olav Dyrkolbotn

Abstract   Digital forensics and, especially, file carving are burdened by the large amounts of data that need to be processed. Attempts to solve this problem include efficient carving algorithms, parallel processing in the cloud and data reduction by filtering uninteresting files. This research addresses the problem by searching for data where it is more likely to be found. This is accomplished by creating a probability map for finding unique data at various logical block addressing positions in storage media. SHA-1 hashes of 512 B sectors are used to represent the data. The results, which are based on a collection of 30 NTFS partitions from computers running Microsoft Windows 7 and later versions, reveal that the mean probability of finding unique hash values at different logical block addressing positions vary between 12% to 41% in an NTFS partition. The probability map can be used by a forensic analyst to prioritize relevant areas in storage media without the need for a working filesystem. It can also be used to increase the efficiency of hash-based carving by dynamically changing the random sampling frequency. The approach contributes to digital forensic processes by enabling them to focus on interesting regions in storage media, increasing the probability of obtaining relevant results faster.

Keywords: File carving, hash-based carving, partition content map, NTFS

1.   Introduction

The ever-increasing amounts of data handled in digital forensic investigations are a major challenge [55]. This situation has been discussed for years [8, 19, 28, 54, 58], but the challenges persist. The research community has attempted to address the problem using a number of approaches. A survey by Quick and Choo [55] lists data mining, data re-

© IFIP International Federation for Information Processing 2019
Published by Springer Nature Switzerland AG 2019
G. Peterson and S. Shenoi (Eds.): Advances in Digital Forensics XV, IFIP AICT 569, pp. 133–158, 2019.
https://doi.org/10.1007/978-3-030-28752-8_8

duction and subsets, triage, intelligence analysis and digital intelligence, distributed and parallel processing, visualization, digital forensics as a service and various artificial intelligence techniques.

File carving is especially affected by the increasing amounts of data. It is used in situations where a filesystem is not present – only the properties of the stored data are available [51, 53]. Previous research has attempted to determine the data type (file type) of fragmented data using histograms of the frequencies of bytes, byte pairs and the differences between consecutive byte values [36–41]. Researchers have also leveraged the compressibility of data for type identification [2–4]. As in previous research, this work uses small blocks of data (512 B sectors) and their statistical properties to improve file carving; however, the focus is on the most probable positions of user data as opposed to their exact types.

Carving files without the help of a working filesystem is difficult, but such a capability is very valuable in digital forensic investigations. In hash-based carving, hashes of blocks of unknown data from storage media are compared against equally-sized blocks of known suspicious material. A number of strategies, techniques and algorithms for hash-based carving have been developed [6, 7, 15, 24–26, 66]. However, the large number of hash comparisons that have to be performed by a hash-based carving algorithm imposes a significant burden on the forensic process.

The research community has not as yet leveraged the principle of searching for data where it is more likely to be found. Since the allocation algorithm of an operating system places new data in a filesystem according to a set of rules (and not randomly), this principle can be used in digital forensics. However, the allocation process is too complex to understand completely and is, therefore, commonly considered to be random. As a result, the conventional approach is to search storage media from beginning to end regardless of the most probable positions of the sought-after data.

Most data of interest in forensic investigations is related to user activity (e.g., system logs and user-created files). Such data is often unique to a specific computer because the probability of two users independently creating exactly the same data is miniscule. Shared data (e.g., child abuse material) downloaded by a user is also of interest. But this data is intertwined with unique user data in storage media according to the rules of the allocation algorithm. Therefore, it is sensible to use the logical block addressing (LBA) positions of unique user data to find data related to user activities.

To demonstrate the principle, this chapter describes an experiment that uses SHA-1 hashes of the content of non-related computers running Windows 7 and later versions to compute the probabilities of unique hashes (corresponding to unique user data) at different positions in the 30 largest NTFS-formatted partitions of 26 hard disks. The data was chosen to be as realistic as possible in order to increase the applicability of the results; for this reason, data from real-world computers was collected for the experiment.

2. Related Work

Although the approach described in this chapter is unique, it is instructive to evaluate related research in the area of file carving.

2.1 File Fragment Carving

Veenman [65] has employed the entropy, histograms and Kolmogorov complexity of 4 KiB file fragments to determine their types; the results reveal that histograms yield the highest detection rate versus false positives. Calhoun and Coles [11] have experimented with statistical measures such as ASCII code frequency, entropy, mode, mean, standard deviation and correlation between adjacent bytes; they have also considered the use of the longest common sub-strings and sub-sequences between file fragments for data classification.

Ahmed et al. [1] have used byte frequency distributions to measure the distances between the statistical properties of a data fragment and a model; instead of using the Mahalanobis distance measure, they employed cosine similarity and obtained better results. Li et al. [43] have also used byte frequency distributions (histograms) of different data fragments, but in conjunction with a support vector machine to discriminate between data types. The best results were achieved using byte frequency distributions on their own.

Fitzgerald et al. [23] have combined several statistical measures of data fragments (histograms of one- and two-byte sequences, entropy and Kolmogorov complexity) to obtain feature vectors that were fed to a support vector machine for classification. Their method outperformed the approaches proposed by other researchers. However, they did not evaluate the contributions of the chosen feature vectors, leaving it for future work. Interested readers are referred to Poisel et al. [52] for a taxonomy of data fragment classification techniques.

2.2 Hash-Based Carving

Hash-based carving compares the hashes of known file blocks against the hashes of equally-sized blocks from storage media. This approach enables files that have been partially overwritten or damaged to be identified.

The roots of hash-based carving go back to the spamsum tool developed by Tridgell [62]. Garfinkel and McCarrin [25] were among the first to use hashes for file carving (specifically, in the 2006 Digital Forensic Research Workshop (DFRWS) Carving Challenge). Subsequently, Kornblum [42] employed spamsum for piecewise hashing, which is now referred to as approximate matching. Dandass et al. [17] have used hashes for file carving in an empirical study of disk sector hashes. However, the term hash-based carving was first introduced by Collange et al. [15] when they explored the possibility of using a graphics processing unit (GPU) to compare hashes of 512 B sections of known files with hashes of 512 B sectors from disk images.

In the 2006 DFRWS Carving Challenge [25], portions of files found on the Internet were hashed and used to find the same hashes in the challenge image. The experience led to the development of the frag_find tool [26]. The optimal size of the data blocks to hash was determined to be equal to the sector size; however, it is not mentioned if the sectors were 512 B or 4 KiB in size. Garfinkel et al. [25] elaborated further on the size of hashed blocks, stating that, starting with Windows NT 4.0, the default minimum allocation unit in NTFS is 4 KiB [48].

Foster [24] discusses the problem of data shared across files, noting that "the block of nulls is the most common block in [the] corpus," relating them to NULL paddings in files. Young et al. [66] have further developed Foster's ideas; they discuss the optimal block size, the handling of large amounts of data, efficient hash algorithms, good datasets to use and common blocks of files.

Random sampling has been used to improve the speed of hash-based carving [24–26]. The determination of a suitable sampling frequency is regarded as sampling without replacement. A higher sampling frequency may increase the detection rate, but it negatively impacts execution speed. The key problem is to strike a trade-off between detection rate and execution speed.

2.3   Data Persistence

The concept of data persistence is relevant to this research because persistence in areas of storage media indicates that these areas have not been reused. This information is valuable when creating a map of storage media.

Jones et al. [35] have created a framework for studying the persistence of deleted files in storage media. They employ differential forensic analysis to compare snapshots of filesystems in use and follow the decay of deleted files over time.

Fairbanks and Garfinkel [22] identify twelve factors that affect data persistence in storage media. Fairbanks [20, 21] also discusses the low-level functions of Ext4 and their impacts on digital forensics.

2.4 Data Reduction

Quick and Choo [54, 56] have proposed methods for reducing the amounts of data analyzed in digital forensic investigations. They extract specific files using a list of key files and work on the subset of files. However, this requires a working filesystem, which limits the applicability of their methods. Also, the list of key files has to be updated constantly.

Rowe [59] has proposed a similar approach to that of Quick and Choo, although it is more technical. Nine methods are compared for identifying uninteresting files, which are defined as "files whose contents do not provide forensically useful information about the users of a drive." However, all the methods require a working filesystem, which is inconsistent with the fundamental premise of file carving.

2.5 Data Mapping

Guidance Software [34] has developed an EnScript module for its En-Case software, which creates a map of the recoverable sectors of a file found in a filesystem. The module can handle situations where other tools do not work (e.g., partially damaged files). However, it is very processor intensive and can create maps of only a few files at a time.

Gladyshev and James [28] have studied file carving from a decision-theoretic point of view. They specify a model where the storage media is sampled at a frequency based on the properties of the hard drive and the file type that is to be found. In some situations, their carving model outperforms standard linear carving algorithms, but their solution is not general. Gladyshev and James also mention using the distribution of data on disk, but do not relate this to the probability of finding user data at various logical block addressing positions in storage media.

Van Baar and colleagues [63, 64] focus on the non-linear extraction of data from images. They posit that master file table (MFT) records (filesystem metadata) of an NTFS partition should be extracted first. The MFT records can be used to find interesting areas in the filesystem. They also leverage the analysis process to influence the imaging process by prioritizing certain portions of storage media during imaging.

3.   Proposed Method

The review of the literature reveals that it is necessary to improve the efficiency of algorithms and tools used in digital forensics, and especially in file carving. However, the research community has not yet leveraged the inherent structures of allocation algorithms to address the problem. This research is founded on the novel idea of using the probability of finding user data in various locations in storage media to govern the digital forensic process and, thus, contribute to an immediate increase of the efficiency of current file carving algorithms and tools. In hash-based carving, the concept can be used to increase the efficiency of random sampling by varying the sampling rate according to the probability of finding user data at different logical block addressing positions. The concept can also be used during triage and other situations that involve trade-offs between speed and detection rate.

Another advantage is that the proposed method works without a filesystem. The map created by the method can be used directly to improve the speed of any file carving algorithm by revealing the most probable positions of unique data in NTFS-formatted storage media. The forensic process can be started at the position with the highest probability of containing data of interest (e.g., user data) or by varying the sampling rate according to the probability of finding user data. In the latter case, the sampling frequency would be higher where it matters most and lower in other areas, increasing the likelihood of getting hits while maintaining the same amount of samples as in the case of equally-distributed sampling.

The proposed method also benefits digital forensic analysts because the map can help plan a forensic process in the same way a road map is used to plan a trip. Storage media are currently treated as black boxes, which forces forensic analysts to spend valuable time scanning them from beginning to end before any analysis can be performed. This is especially true in a file carving situation where there is no filesystem to govern the search. With the proposed method, a forensic analyst can focus on relevant areas of the storage media and postpone, or even skip, less relevant areas.

The map can also support storage media imaging. By starting the imaging process at the most probable position of user data and continuing in decreasing order of relevance, the analysis process can be run almost in parallel with imaging because the most relevant data for analysis is available immediately. Thus, the analysis process can be started earlier, even before the imaging is finalized, saving valuable time and effort. Of course, the reliability of the analysis will increase as more data is analyzed, but preliminary results are available that can guide the subsequent work. This concept is also supported by the Hansken Project [63, 64]. By implementing the proposed method in Hansken, its ability to handle media with broken filesystems would be enhanced, possibly approaching the performance of the standard process.

The proposed method uses 512 B sectors when hashing data in order to handle storage media regardless of its filesystem cluster size. Since the map is created once and can be reused, as in the case of a road map, no performance penalty is incurred when using the map. The map is divided into a small number of equally-sized areas (currently 128) so that any random seek penalty would only occur between – and not within – these areas, and could, therefore, be ignored.

The only situation where 512 B hashes are required is when the map is created. There is no need to use 512 B hashes when performing casework. Likewise, any hashing algorithm can be used in casework because the map hashes are only used to compute the probabilities of user data at different positions, never for comparing hashes from a specific case. If a given hash algorithm is broken, it could be replaced by a new algorithm and the map would still work.

During the research, nine sectors were discovered to have the same hash values at the same logical block addressing positions in all 30 partitions. These sectors could be used to identify an NTFS filesystem, to find the start of an NTFS partition and to locate the $MFT file for further processing. These tasks can be accomplished regardless of the state of the filesystem.

4. Experimental Setup

Live data was collected from real computers in order to determine the distribution of unique data in the major NTFS-formatted partitions of common Microsoft Windows computers. Next, the probabilities of finding unique hash values at different logical block addressing positions were computed. Finally, a map was created by computing the mean probabilities of a number of (currently 128) equally-sized partition areas based on logical block addressing positions. The mean probability computations were performed to generalize and scale the map into a usable format.

In order to reduce the size of the data stored in the experiment and to protect the privacy of users, the SHA-1 algorithm was used to hash each 512 B sector of all 30 NTFS-formatted main partitions considered in the experiment. SHA-1 was employed because it currently strikes the best balance between speed, collision risk and hash size from among the candidate hash algorithms (MD5 and SHA family). The choice was based on a practical evaluation using the available hardware. The hashing was performed locally at each source computer and only the resulting hashes were moved from the source computers.

Since the SHA-1 algorithm maps 512 B of data to a 20-byte hash, there is always a risk of collisions. The collision risk arises because 512 B of data is compressed into a 20-byte hash and, therefore, the results may contain false positives. The problem can be analyzed using the Birthday Paradox, where N is the number of possible hashes, n is the number of hashes (i.e., total number of sectors hashed in the worst case) and $Prob$(Collision) is the probability of a collision, which is given by:

$$Prob(\text{Collision}) = 1 - \frac{N!}{N^n \cdot (N - n)!}$$

When $N = 2^{160}$ and $n = 18,210,308,798$, the probability of at least one collision is:

$$Prob(\text{Collision}) \approx 1 - e^{-n^2/2N} \approx n^2/2N \approx 1.1 \cdot 10^{-28}$$

This expression uses Stirling's approximation of factorials, which yields acceptable results for large numbers.

Since the SHAttered attack [16, 61] is 100,000 times faster than a brute force attack using the Birthday Paradox, the risk of an intentional collision is higher. Fortunately, the attack is infeasible in digital forensic scenarios. Therefore, a unique SHA-1 hash is assumed to represent a unique piece of data.

Although the SHA-1 algorithm has been broken [16, 61], from a cryptographic point of view, the risk of an intentional collision is also negligible. This is because the computing power required to create a collision is out of reach of common users. Also, such an attack would require a large number of collisions to be created for the majority of the storage media in the map source data. In a digital forensic scenario, it would be much easier to fill the storage media with shared and unique data in an intentional pattern; but this is easily mitigated by collecting source data from unrelated sources. In any case, the mapping process is not limited to SHA-1 – any hashing algorithm would suffice as long as all the mapping data is hashed using the same algorithm.

4.1   Data Collection

Data representing real-world situations was obtained by using convenience sampling of data from computers owned by acquaintances. The Real Data Corpus was not used because the timestamps on the website [18] indicate that the last update to the dataset was made back in

2011. The data collected for the experiment was much newer, corresponding to Windows versions 8 and 10. Windows 8 was introduced at the end of 2012 [27] and, therefore, Windows 8 data does not exist in the Real Data Corpus, nor does Windows 10 data.

Data was collected from 30 partitions in 26 computers (23 consumer-grade and three office-grade computers). The data was collected by hashing every 512 B sector of the drives using the dcfldd disk imaging tool set to the SHA-1 hash algorithm. The operating system installations corresponded to three language packs ranging from Microsoft Windows 7 to Windows 10 Enterprise, Professional, Ultimate, Home and Educational versions. Some computers have been upgraded from an earlier Windows version to Windows 10. However, five computers are maintained in their original form and access to all their raw content is still available.

Real computers were employed to avoid bias introduced by simulating user behavior. While real computers ensure that the results would be as close to possible to casework, the drawback was reduced control of the material. For example, in some instances, information whether a drive was mechanical or solid-state was not available. The lack of information does not affect the results because data was collected at the logical block addressing level from the drive controller. However, the lower level physical storage formats were hidden [5, 10, 14, 57].

As far as the experiment is concerned, the only difference between a mechanical hard disk and a solid state drive is how the unused areas were filled. This could be old data, 0x00 or 0xFF depending on how the TRIM command was implemented in the solid state drives [9, 29–33]. Hence, a mechanical drive would more often yield old data from unallocated clusters compared with a solid state drive. Since only the logical block addressing positions of unique data was used, any 0x00 and 0xFF fillings were automatically filtered. In the case of old data from unallocated clusters, a very unbalanced erase/write cycle would leave a large amount of old data, i.e., a large amount of data would have been erased first, followed by a small amount of (or no) writing of new data. This would be the case if a hard disk was erased using a random pattern and then reformatted and reused. The experimental results would be affected if a large number of unallocated sectors were to contain old (unique) data. In order to affect the map creation process to a sufficient degree, this situation would have to hold for a significant number of partitions in the dataset. Therefore, data was not collected from the computers until their users confirmed that large-scale filesystem cleaning was not performed close to the data collection.

Table 1. Partition sizes, unique hashes, and 0x00 and 0xFF fillings in the last 20 GB.

Partition	Size (GiB)	Unique Hashes (%)	0x00 Filling (%)	0xFF Filling (%)
F	59.5	0.08	100.00	0.00
E	59.5	22.36	2.01	0.12
AC	111.3	7.63	100.00	0.00
I	111.6	61.83	20.12	1.67
A	118.4	23.17	75.24	0.00
W	118.6	59.80	26.18	0.00
K	146.4	5.82	100.00	0.00
Qa	150.0	43.33	45.78	0.07
N	177.6	38.32	48.48	0.00
Ra	185.9	31.89	56.89	0.14
Sa	200.0	86.85	0.02	0.02
Oa	209.0	13.68	100.00	0.00
Y	217.1	14.77	100.00	0.00
P	232.7	53.97	0.83	0.07
H	237.3	16.68	100.00	0.00
AA	237.3	12.60	100.00	0.00
D	237.9	20.59	0.00	100.00
G	238.1	7.03	25.56	0.16
M	238.1	23.06	79.47	0.01
Rb	258.4	1.68	100.00	0.00
Sb	265.6	36.22	100.00	0.00
T	297.9	9.35	48.12	0.28
C	421.7	34.98	0.86	1.17
Z	423.9	4.05	100.00	0.00
X	443.8	6.48	100.00	0.00
U	448.0	10.60	100.00	0.00
V	465.6	48.43	100.00	0.00
Ob	699.0	0.15	98.72	0.00
Qb	766.5	1.47	100.00	0.00
B	905.2	29.74	100.00	0.00
Total	8,683.4	20.92	67.61	3.35

The storage media used in the experiment had capacities ranged from 64 GB to 1 TB. The largest NTFS-formatted partition was extracted from each storage media drive based on the assumption that it contained the operating system and user files. In four instances, an extra storage partition was present, which was also extracted.

Table 1 shows the partition sizes, percentages of unique hashes and the 0x00 and 0xFF fillings in the last 20 GB of the partitions. The total size of the partitions was 8,638.4 GiB, corresponding to 18,210,308,798 hashes, 3,809,786,792 of which were unique. A low percentage of unique hashes in a partition indicates that the partition was not used or at least not used for storing user data. The larger partitions in Table 1 (i.e., those ending in "b") have low percentages of unique hashes and high percentages of 0x00 and 0xFF fillings. Data was collected from multiple partitions on the associated drives, which were large.

The lifetimes and, hence, amounts of data stored on the drives varied. Most of the drives had 0x00 fillings to some extent. These may be remnants of the production process, but some of the smaller drives ($\leq$ 256 GiB) were solid state drives that would have been filled with 0xFF values at the factory [9].

The last 20 B of each partition was examined to determine if any partitions in the dataset were completely filled with data at some point in their lifetimes. The 20 GB size was selected as a suitable trade-off between having a large amount of data and the risk of including operating system files in the case of the smaller partitions. Low percentages of both 0x00 and 0xFF fillings in Table 1 is an indicator that the partition was completely filled or was wiped with a random pattern at some point in its lifetime.

NTSF-formatted partitions were considered because NTFS has about 90% of the desktop system market share [50]. The partition names in Table 1 were assigned based on the order of hashing (i.e., partition "A" was hashed before "B"). Four computers contributed two partitions each to the dataset due to the partition size (these computers were installed with an extra partition for user data). The associated partitions are indicated by a second lowercase letter in their names. Although the partitions did not have operating system files, they contained an NTFS filesystem and were, therefore, included in the dataset.

Some of the unique hash values that were found corresponded to 1 KiB MFT records. The size of an MFT record is defined in the boot sector of an NTFS partition; the standard size is 1 KiB [12]. These records yielded up to two unique hash values each due to their highly varying content (timestamps, file names, file content, etc.).

Therefore, a survey of 27 computers not included in the dataset was conducted to estimate the mean number of MFT records. This was accomplished by counting the total number of files and folders because each file and folder in a computer is represented by at least one MFT record. If a file has many attributes (e.g., alternate streams) or is heavily fragmented, then the filesystem creates a new MFT record to hold the extra information.

The survey revealed that the average total number of files and folders in the computers was 363,630. Due to the uncertainty involved in counting (using File Explorer), an extra 25% was added to account for hidden files, files requiring more than one MFT record and MFT records internal to the filesystem. The extra 25% also covered network storage of user data in the office-grade computers. In the case of the consumer-grade computers, all the user files would most likely have been stored locally; therefore, these files were included in the count.

4.2   Implementation

In order to prepare the data for the experiment, the hashes from the largest partitions were computed and merged into a single file, which was then sorted in ascending order of hash value. The unique hashes from the file were then extracted; thus, all the 0x00- and 0xFF-filled sectors were automatically filtered out. The unique hashes were then sorted in order of ascending logical block addressing position and separated into individual files based on partition. The data in each partition was then divided into 128 equally-sized areas, each $\frac{1}{128}$ of the size of the partition. Following this, the probabilities of finding unique hashes in each area were computed as the number of unique hashes divided by the sizes of the areas in the sectors in each partition.

After the probability computations, the mean, median and standard deviation of the probabilities of unique hashes were calculated for each area of the partitions regardless of the partition sizes. The mean values were used to create a map of the storage media, showing where it is more likely to find user data (unique data) in a generic-NTFS formatted partition.

4.3   Evaluation

The map was evaluated by conducting an experiment that simulated a hash-based carving scenario and comparing the performance of sampling based on the map against that of a uniformly-distributed sampling. Four real partitions that were not included in the dataset were used as ground truth. The distributions of unique data in the four partitions were used to pick a random integer *target*. The map was then used to pick a random integer *map* and the uniform distribution was used to pick a random integer *uni*. All the random integers were selected in the same total range representing the logical block addressing positions of a fictitious partition, albeit with a bias for *target* and *map*. The map received one hit when *map* = *target* and the uniform distribution received one hit when *uni* = *target*. The predefined range was set to 16 MiB and divided into 128 equally-sized areas using the mapping process. The small partition size was chosen to increase the number of hits.

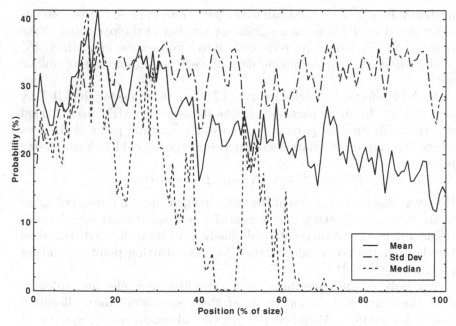

Figure 1. Probabilities of unique hashes.

The random sampling process was iterated 10^9 times for each of the four partitions to stabilize the results. The small number of partitions used to create the map adversely affected the evaluation results. Since the number of partitions used to create the ground truth was also small, the results were affected by individual variations in partition content. Another factor affecting the results is that the four partitions used as ground truth were taken from computers that were about to be scrapped; since their drives were well-used, the partitions had lower percentages of 0x00 and 0xFF fillings towards their ends.

The experiment used Python 2.7 and the `random` library running on a Debian Stretch (version 9) computer.

5. Experimental Results

Figure 1 shows plots of the means, medians and standard deviations of the probabilities of unique hashes at different positions in the 30 partitions in the dataset. The positions are presented as percentages of the partition size. Each partition was split into 128 equally-sized areas based on the total size of the partition.

As seen in the map in Figure 1, the probabilities of unique hashes at different positions vary from approximately 12% to 41%. The low median values in the second half of the partitions are due to the presence of 0x00 and 0xFF fillings in a significant number of the partitions. Note that the median values are zero or close to zero because more than 50% of the partitions had no unique data or very small amounts of unique data.

An NTFS-formatted drive reserves 12.5% of its space for the MFT by default [49]. In all 30 partitions in the dataset, the MFT area started exactly 3 GiB into the partitions. Hence, the starting point of the area where non-resident file data is allocated is at the logical block addressing position:

$$P = 3 \cdot 2^{30} + 0.125 \cdot partition\,size \text{ (bytes)}$$

However, this is not the case if the user changed the MFT reserved space at the time of formatting. In the case of a very small partition, the non-resident data allocation point would likely have been changed. Based on the dataset, the non-resident data allocation starting point is valid for partitions $\geq$ 60 GiB.

The bulk of the operating system, the first user files and software from the initial installation reside at the non-resident data allocation point. According to Microsoft [45–47], the minimum space requirement for Windows 7, 8, 8.1 or 10 installations is 20 GiB for 64-bit systems. Thus, the most probable starting point for day-to-day usage of a partition containing Windows is:

$$(3 + 20) \cdot 2^{30} + 0.125 \cdot partition\,size \text{ (bytes)}$$

bytes into a partition. Converting the starting points to percentages of the associated partition lengths yields values between 14% and 43% (see Figure 1).

The largest number of operating system files were found in the beginning of the area and the number decreased towards the end. This explains the sharp negative trends in the plots between 20% and 40%, along with the peaks around 30%. In the case of the mean plot, from 40% onwards, the values slowly decrease and the standard deviations increase. This is due to differing usage patterns of the partitions – some partitions had been storing more data and/or had been more utilized than other partitions in the dataset.

A total of 3,809,786,792 unique hashes were found in the dataset; these correspond to data created locally by the user or operating system (e.g., logs). However, unique portions of MFT records were also in the data. Each file and directory is represented by at least one MFT record in NTFS. Depending on the number of file attributes, multiple MFT records may be needed to store the metadata. Typical examples are files with many alternate data streams or highly fragmented files.

```
Type: $DATA (128-12)   Name: N/A   Non-Resident   [...]
786432 786433 786434 786435 786436 786437 786438 786439
[...]
Type: $DATA (128-1)   Name: N/A   Non-Resident   [...]
786432 786433 786434 786435 786436 786437 786438 786439
[...]
Type: $DATA (128-6)   Name: N/A   Non-Resident   [...]
786432 786433 786434 786435 786436 786437 786438 786439
[...]
Type: $DATA (128-1)   Name: N/A   Non-Resident   [...]
786432 786433 786434 786435 786436 786437 786438 786439
[...]
Type: $DATA (128-6)   Name: N/A   Non-Resident   [...]
786432 786433 786434 786435 786436 786437 786438 786439
[...]
```

Figure 2. Portions of the $DATA attributes of $MFT files

MFT records may affect the results by increasing the number of unique non-user data hashes. To estimate the effect of the MFT records, the numbers of files and folders in 27 typical computers (office-grade and home-grade) were examined. The mean value was found to be 363,630 files, which corresponds to approximately 0.7% of the unique hashes in the 30 computers.

The `pagefile.sys` and `hiberfil.sys` files may also generate large numbers of unique hashes depending on how much they were used. These files certainly affect the map and the results. However, they should be included because they are of high value in a digital forensic investigation.

When studying the mapping process, four sectors were found to contain the same hash value at the same logical block addressing position in all the partitions in the dataset. The sectors, located in filesystem cluster 786,435, all contained the second half of MFT records that had been used only once according to their signature values [12]. The first part of the MFT records contained similar, but not equal information. The `istat` tool [13] revealed that the sectors belonged to the $MFT file (i.e., filesystem itself). The $DATA attributes of the $MFT files in the five computers to which raw access was still available allocated the same eight clusters at the beginning of the run length. Note that the eight numbers in every third row in Figure 2 indicate the filesystem clusters allocated to a file. Filesystem cluster 786,435 contained four static sectors – at positions 6,291,481, 6,291,483, 6,291,485 and 6,291 487 – that were found in all 30 partitions.

Table 2.   Evaluation results.

Hits (Map)	Hits (Uniform)	Percentage (Map/Uniform)
28,635	30,279	94.6%
29,881	30,363	98.4%
32,556	30,836	105.6%
33,257	30,461	109.2%
124,329	121,939	102.0%

Upon combining this with the static content of the four sectors in cluster 786,435, the NTFS formatting appears to place the start of the MFT at the same position – exactly 3 GiB into the partition. If this is true, then the first and last sectors of an NTFS partition should contain the hex string 00 00 0C 00 00 00 00 00 starting at position 0x30 (little endian) [44]. This was verified for the five computers for which raw access was still available.

According to Carrier [12], the $DATA attribute of the $MFTMirr file allocates clusters in the middle of a filesystem. This implies that the middle sector, based on the size of the volume (partition), is actually where the mirror should be kept. However, this was not always true. In four of the five computers to which full access was available, the $MFTMirr file allocated filesystem cluster 2 and, in the last partition, filesystem cluster 8,912,895 was allocated. However, the latter partition was 59,919,808 clusters in size; hence, none of the $MFTMirr files were located near the middle of any of the partitions. Consequently the NTFS allocation strategy appears to have changed since Carrier published his book [12].

In order to evaluate the efficiency of the map in random sampling situations, four NTFS partitions (different from the 30 partitions in the mapping dataset) were used to create the map. The four 16 MiB partitions were divided into 128 equally-sized areas that were sampled 10^9 times in the evaluation.

Table 2 shows the evaluation results. In particular, the table shows the number of hits using the map relative to using a uniformly-distributed sampling rate. Due to the small number of partitions used in the evaluation, the distributions of unique data in the individual partitions have a high impact on the result. Therefore, the evaluation results are merely the first indicator of the performance of future maps, not the final answer (future research will conduct a new evaluation using a larger number of partitions). In any case, the best result – when the map most resembles one of the evaluation partitions – is almost 10% better than using a uniformly-distributed sampling rate. Changing the number of equally-sized areas does not change the results in any significant way. Changing the 16 MiB partition size also does not change the results.

6. Discussion

The concept of searching for data where it is more likely to be found is more appealing than randomly searching for data. Nevertheless, an empirical evaluation was conducted to test the implementation. The evaluation reveals a 2% improvement when using the map compared with using a uniformly-distributed sampling rate. This is certainly not a paradigm shift, but it is a positive indicator of the utility of the concept.

The reason for the modest result is the small number of partitions used to create the map. More partitions are required to reveal the underlying deterministic allocation pattern; this would also provide a solid statistical foundation and improve the strength of the results. Having a large enough dataset would enable its division into several use cases, each producing its own map. An example would be to differentiate between web surfers, office administrators and file sharers. However, such a study would require a much larger data collection effort while maintaining a high level of control of the collected material to filter unique data that was not created by the user or system (e.g., data written during disk wiping).

After the mapping foundation is stable, the efficiency of digital forensic methods and tools – especially in file carving situations where a filesystem is not present – can be improved in several ways. One example is when using hash-based carving to find portions of files on storage media. In this case, the following three strategies are possible:

- **Speed is Prioritized:** The total number of samples is reduced compared with the uniformly-distributed sampling case without any significant loss in detection rate. This strategy could be used in triage situations or when preliminary results are required quickly.

- **Speed is Maintained:** The same amount of samples are maintained compared with the uniformly-distributed sampling case, which enhances detection at the same execution speed. This standard case can be used without changing the digital forensic process.

- **Detection Rate is Prioritized:** A larger number of samples are used than in the case of uniformly-distributed sampling, yielding a much higher detection rate and a lower cost in terms of execution speed. An example is a situation where a suspect's drive has an unusual usage pattern. The standard number of hashes can be maintained in low priority areas of a drive whereas a higher sampling rate can be used in high priority areas for improved detection.

When the area reserved for the MFT is used up, a new area amounting to 12.5% of the volume size is added. This area may be contiguous, but it does not have to be so. As the filesystem grows, new MFT records are added and allocated where suitable [12]. Thus, an old and well-used NTFS partition may very well have MFT records spread all over the storage space. This would possibly affect the creation of the map, adding noise to the unique data.

According to the empirical study of the number of files and directories (usually represented by a single MFT record each) in an NTFS partition, the amount of MFT records corresponds to approximately 0.7% of the total amount of unique hashes in each partition. The actual amount of unique hashes belonging to an MFT record is probably less than 0.7% because the second part of an MFT record often contains 510 zero-bytes followed by a two-byte signature value at the end of the sector. Signature values are used by NTFS to verify the integrity of data structures (but not sectors containing file content) that span two or more sectors [12]. The last two bytes of every sector in such a data structure are called a fixup value and are moved to an array in the beginning of the structure during the process of writing to disk. The last two bytes are then replaced by the signature value. When the data structure is read, the signature values are used to check that all the sectors that are read have the same signature value and, thus, belong to the same data structure. The signature value is incremented by one every time a data structure is updated on disk [12].

The worst case scenario is a partition filled with files that are less than approximately 700 B in size, resulting in a partition filled with MFT records that store the data internally. The maximum size of an internal $Data attribute varies depending on the sizes of other attributes stored in the MFT record. Most sources give a maximum internal $Data attribute size of 600 to 700 bytes; Microsoft reports a 900 B limit [49]. If all the files contain the same data, only the MFT metadata (timestamps, etc.) would differ; thus, the partition would still appear to be filled with random data. The maximum number of files in an NTFS partition is $2^{32} - 1$ [49], so the partition would be approximately 4 TiB in size.

Another way to estimate the number of unique MFT record hashes in the dataset is to generate SHA-1 hash values for all possible combinations of 510 zeros and a two-byte signature value, which correspond to the second half of a standard MFT record. The first such hash that is unique in the dataset represents a signature value of 3613 (0x1D0E little endian). Many of the lower signature values generate several thousand hits. There is no guarantee that all the generated hashes belong to MFT records; however, at least four do and, consequently, the percentage of unique MFT hash values that pollute the dataset is likely less than 0.7%. Hence, the unique hashes of the MFT records do not significantly affect the precision of the map.

The experiment was limited to computers running Microsoft Windows 7 and later versions with NTFS-formatted main partitions. The privacy of the computer owners was protected by using SHA-1 hashes to obscure the real data. This limited the ability to trace the original data corresponding to each hash. However, since the focus is on the logical block addressing positions of unique hashes, the precise data represented by the hashes is not needed for map creation.

The proposed method can also be used to find shared data. Of special interest is static data – shared data found in the same logical block addressing position in unrelated storage media. Knowledge of the logical block addressing positions of static data is of value in several digital forensic applications. For example, prioritizing search in forensic imaging and analysis could provide an analyst with the means to break drive encryption via a plaintext attack [60] (depending, of course, on the encryption algorithm that is used).

The logical block addressing position of static data can be used to handle corrupt storage media. In many cases, large portions of corrupt media are readable, but there are no indications of the forensic value of the lost portions. Having access to a map of static content in storage media can help a digital forensic analyst improve the evaluative reporting in casework by indicating the forensic value of lost areas. This contributes to higher confidence in the collected evidence.

Furthermore, a map can be used to create signatures that identify the filesystems in partially-recovered partitions. These signatures are feasible because the metadata layout and allocation process during installation differ for operating systems.

Finally, media areas that should have high probabilities of static content but do not have this content could be indicators of the presence of malware or other suspicious activities because deviations are unlikely in such areas. Instead of having to hash every file in a filesystem in search of deviations, the search can start at the most likely place in the filesystem. The partition is then scanned in descending order of probability of static content.

7.   Conclusions

The proposed method for hash-based file carving is based on the principle that it is better to search for data in locations where the probability of finding the data is high. The method relies on a map of the probability of finding unique data at various logical block addressing positions in storage media. Unique data is data that is created locally on a computer and not (yet) shared. This includes system-created data such as log files and user-created local data that is not downloaded from the Internet (downloaded data corresponds to shared data). Uniquely-created data is often more valuable in a forensic investigation than shared data, although shared data (e.g., child abuse material) can be valuable too.

The map provides a digital forensic analyst with a pre-computed view of storage media that enables the forensic process to focus on the most relevant portions of media instead of spending valuable time scanning the entire storage media from beginning to end. Unique data is only used to create the map; this is done once, although regular updates to the map are recommended. After the map is created, it can be used repeatedly for any data or with any method, tool or investigative process without having to recreate the map for every new case.

Creating a probability map of unique (or static) data at different positions in storage media opens up a host of applications. The map could be used when performing triage, planning the order of analysis of large amounts of seized storage media, estimating the value of partially-analyzed data due to corruption and breaking the encryption of storage media.

Future research will extend the dataset to stabilize map creation and make the map more reliable. It will also explore other methods for creating maps as well as creating separate maps for use cases. Research will also search for and study the origin of interesting areas in storage media such as the four sectors with the same hash values found at approximately 3 GiB into all 30 partitions, which indicates that certain areas of NTFS partitions are static. Finally, future work will extend the method to other filesystems, especially Ext4 and Apple File System (APFS), with the goal of creating a general mapping process for storage media, regardless of type and filesystem.

The authors of this chapter are interested in releasing the current hash dataset to the public. However, due to its size, the optimal transfer option will have to be discussed with the interested party. Interested parties are encouraged to contact the first author (`martin@filecarving.net`) to arrange for file transfers.

Acknowledgement

This research was sponsored by the Norwegian Research Council Ars Forensica Project No. 248094/O70.

References

[1] I. Ahmed, K. Lhee, H. Shin and M. Hong, On improving the accuracy and performance of content-based file type identification, *Proceedings of the Fourteenth Australasian Conference on Information Security and Privacy*, pp. 44–59, 2009.

[2] S. Axelsson, The normalized compression distance as a file fragment classifier, *Digital Investigation*, vol. 7(S), pp. S24–S31, 2010.

[3] S. Axelsson, Using normalized compression distance for classifying file fragments, *Proceedings of the International Conference on Availability, Reliability and Security*, pp. 641–646, 2010.

[4] S. Axelsson, K. Bajwa and M. Srikanth, File fragment analysis using normalized compression distance, in *Advances in Digital Forensics IX*, G. Peterson and S. Shenoi (Eds.), Springer, Berlin Heidelberg, Germany, pp. 171–182, 2013.

[5] J. Barbara, Solid state drives: Part 5, *Forensic Magazine*, vol. 11(1), pp. 30–31, 2014.

[6] F. Breitinger and K. Petrov, Reducing the time required for hashing operations, in *Advances in Digital Forensics IX*, G. Peterson and S. Shenoi (Eds.), Springer, Berlin Heidelberg, Germany, pp. 101–117, 2013.

[7] F. Breitinger, C. Rathgeb and H. Baier, An efficient similarity digests database lookup – A logarithmic divide and conquer approach, *Journal of Digital Forensics, Security and Law*, vol. 9(2), pp. 155–166, 2014.

[8] F. Breitinger, G. Stivaktakis and H. Baier, FRASH: A framework to test algorithms of similarity hashing, *Digital Investigation*, vol. 10(S), pp. S50–S58, 2013.

[9] C. Buckel, Understanding Flash: Blocks, Pages and Program Erases, *flashdba Blog* (flashdba.com/2014/06/20/understanding-flash-blocks-pages-and-program-erases), June 20, 2014.

[10] C. Buckel, Understanding Flash: The Flash Translation Layer, *flashdba Blog* (flashdba.com/2014/09/17/understanding-flash-the-flash-translation-layer), September 17, 2014.

[11] W. Calhoun and D. Coles, Predicting the types of file fragments, *Digital Investigation*, vol. 5(S), pp. S14–S20, 2008.

[12] B. Carrier, *File System Forensic Analysis*, Pearson Education, Upper Saddle River, New Jersey, 2005.

[13] B. Carrier, TSK Tool Overview (`wiki.sleuthkit.org/index.php?title=TSK_Tool_Overview`), January 13, 2014.

[14] T. Chung, D. Park, S. Park, D. Lee, S. Lee and H. Song, A survey of the flash translation layer, *Journal of Systems Architecture*, vol. 55(5-6), pp. 332–343, 2009.

[15] S. Collange, Y. Dandass, M. Daumas and D. Defour, Using graphics processors for parallelizing hash-based data carving, *Proceedings of the Forty-Second Hawaii International Conference on System Sciences*, 2009.

[16] Cryptology Group at Centrum Wiskunde and Informatica and Security, Privacy and Anti-Abuse Group at Google Research, SHAttered – We have Broken SHA-1 in Practice (`shattered.io`), 2017.

[17] Y. Dandass, N. Necaise and S. Thomas, An empirical analysis of disk sector hashes for data carving, *Journal of Digital Forensic Practice*, vol. 2(2), pp. 95–104, 2008.

[18] Digital Corpora, Real Data Corpus (`digitalcorpora.org/corpora/disk-images/real-data-corpus`), July 15, 2018.

[19] EUROPOL: European Law Enforcement Agency, IOCTA 2016: Internet Organized Crime Threat Assessment, Technical Report, European Police Office, The Hague, The Netherlands, 2016.

[20] K. Fairbanks, An analysis of Ext4 for digital forensics, *Digital Investigation*, vol. 9(S), pp. S118–S130, 2012.

[21] K. Fairbanks, A technique for measuring data persistence using the Ext4 file system journal, *Proceedings of the Thirty-Ninth Annual IEEE Computer Software and Applications Conference*, vol. 3, pp. 18–23, 2015.

[22] K. Fairbanks and S. Garfinkel, Column: Factors affecting data decay, *Journal of Digital Forensics, Security and Law*, vol. 7(2), pp. 7–10, 2012.

[23] S. Fitzgerald, G. Mathews, C. Morris and O. Zhulyn, Using NLP techniques for file fragment classification, *Digital Investigation*, vol. 9(S), pp. S44–S49, 2012.

[24] K. Foster, Using Distinct Sectors in Media Sampling and Full Media Analysis to Detect Presence of Documents from a Corpus, Master's Thesis, Department of Computer Science, Naval Postgraduate School, Monterey, California, 2012.

[25] S. Garfinkel and M. McCarrin, Hash-based carving: Searching media for complete files and file fragments with sector hashing and hashdb, *Digital Investigation*, vol. 14(S1), pp. S95–S105, 2015.

[26] S. Garfinkel, A. Nelson, D. White and V. Roussev, Using purpose-built functions and block hashes to enable small block and sub-file forensics, *Digital Investigation*, vol. 7(S), pp. S13–S23, 2010.

[27] S. Gibbs, From Windows 1 to Windows 10: 29 years of Windows evolution, *The Guardian*, October 2, 2014.

[28] P. Gladyshev and J. James, Decision-theoretic file carving, *Digital Investigation*, vol. 22, pp. 46–61, 2017.

[29] Y. Gubanov and O. Afonin, Why SSD drives destroy court evidence and what can be done about it, *Forensic Focus*, October 23, 2012.

[30] Y. Gubanov and O. Afonin, Recovering evidence from SSD drives in 2014: Understanding trim, garbage collection and exclusions, *Forensic Focus*, September 23, 2014.

[31] Y. Gubanov and O. Afonin, SSD and eMMC forensics 2016, *Forensic Focus*, April 20, 2016.

[32] Y. Gubanov and O. Afonin, SSD and eMMC forensics 2016 – Part 2, *Forensic Focus*, May 4, 2016.

[33] Y. Gubanov and O. Afonin, SSD and eMMC forensics 2016 – Part 3, *Forensic Focus*, June 7, 2016.

[34] Guidance Software, File Block Hash Map Analysis, Version 8.8.5, Waterloo, Canada (www.guidancesoftware.com/app/File-Block-Hash-Map-Analysis), 2018.

[35] J. Jones, T. Khan, K. Laskey, A. Nelson, M. Laamanen and D. White, Inferring previously uninstalled applications from residual partial artifacts, *Proceedings of the Eleventh Annual Conference on Digital Forensics, Security and Law*, pp. 113–130, 2016.

[36] M. Karresand, Completing the Picture – Fragments and Back Again, Licentiate Thesis, Institute of Technology: Faculty of Science and Engineering, Linkoping University, Linkoping, Sweden, 2008.

[37] M. Karresand and N. Shahmehri, File type identification of data fragments by their binary structure, *Proceedings of the Seventh Annual IEEE SMC Information Assurance Workshop*, pp. 140–147, 2006.

[38] M. Karresand and N. Shahmehri, Oscar – File type and camera identification using the structure of binary data fragments, *Proceedings of the First Conference on Advances in Computer Security and Forensics*, pp. 11–20, 2006.

[39] M. Karresand and N. Shahmehri, Oscar – File type identification of binary data in disk clusters and RAM pages, *Proceedings of the Thirty-First IFIP TC-11 International Information Security Conference*, pp. 413–424, 2006.

[40] M. Karresand and N. Shahmehri, Oscar – Using byte pairs to find the file type and camera make of data fragments, *Proceedings of the Second European Conference on Computer Network Defense*, pp. 85–94, 2007.

[41] M. Karresand and N. Shahmehri, Reassembly of fragmented JPEG images containing restart markers, *Proceedings of the Fourth European Conference on Computer Network Defense*, pp. 25–32, 2008.

[42] J. Kornblum, Identifying almost identical files using context triggered piecewise hashing, *Digital Investigation*, vol. 3(S), pp. S91–S97, 2006.

[43] Q. Li, A. Ong, P. Suganthan and V. Thing, A novel support vector machine approach to high entropy data fragment classification, *Proceedings of the South African Information Security Multi-Conference*, pp. 236–247, 2010.

[44] LSoft Technologies, NTFS Partition Boot Sector, Mississauga, Canada (`www.ntfs.com/ntfs-partition-boot-sector.htm`), 2018.

[45] Microsoft, Windows 7 System Requirements, Redmond, Washington (`support.microsoft.com/en-us/help/10737/windows-7-system-requirements`), April 12, 2017.

[46] Microsoft, Windows 8.1 System Requirements, Redmond, Washington (`support.microsoft.com/en-gb/help/12660/windows-8-system-requirements`), April 12, 2017.

[47] Microsoft, Windows 10 System Requirements, Redmond, Washington (`support.microsoft.com/en-us/help/4028142/windows-windows-10-system-requirements`), November 20, 2017.

[48] Microsoft, Default Cluster Size for NTFS, FAT and exFAT, Redmond, Washington (`support.microsoft.com/en-us/help/140365/default-cluster-size-for-ntfs--fat--and-exfat`), April 17, 2018.

[49] Microsoft, How NTFS Works, Redmond, Washington (`technet.microsoft.com/pt-pt/library/cc781134(v=ws.10).aspx`), October 28, 2018.

[50] Net Applications, Desktop Operating System Market Share, Aliso Viejo, California (`www.netmarketshare.com/operating-system-market-share.aspx?qprid=10&qpcustomd=0`), 2017.

[51] A. Pal and N. Memon, The evolution of file carving, *IEEE Signal Processing*, vol. 26(2), pp. 59–71, 2009.

[52] R. Poisel, M. Rybnicek and S. Tjoa, Taxonomy of data fragment classification techniques, in *Digital Forensics and Cyber Crime*, P. Gladyshev, A. Marrington and I. Baggili (Eds.), Springer, Cham, Switzerland, pp. 67–85, 2014.

[53] R. Poisel and S. Tjoa, A comprehensive literature review of file carving, *Proceedings of the International Conference on Availability, Reliability and Security*, pp. 475–484, 2013.

[54] D. Quick and K. Choo, Data reduction and data mining framework for digital forensic evidence: Storage, intelligence, review and archive, *Trends and Issues in Crime and Criminal Justice*, no. 480, pp. 1–11, September 2014.

[55] D. Quick and K. Choo, Impacts of increasing volume of digital forensic data: A survey and future research challenges, *Digital Investigation*, vol. 11(4), pp. 273–294, 2014.

[56] D. Quick and K. Choo, Big forensic data reduction: Digital forensic images and electronic evidence, *Cluster Computing*, vol. 19(2), pp. 723–740, 2016.

[57] R. Reiter, T. Swatosh, P. Hempstead and M. Hicken, Accessing logical-to-physical address translation data for solid state disks, U.S. Patent No. 8898371, November 25, 2014.

[58] V. Roussev, Managing terabyte-scale investigations with similarity digests, in *Advances in Digital Forensics VIII*, G. Peterson and S. Shenoi (Eds.), Springer, Berlin Heidelberg, Germany, pp. 19–34, 2012.

[59] N. Rowe, Identifying forensically uninteresting files using a large corpus, in *Digital Forensics and Cyber Crime*, P. Gladyshev, A. Marrington and I. Baggili (Eds.), Springer, Cham, Switzerland, pp. 86–101, 2014.

[60] B. Schneier, *Applied Cryptography: Protocols, Algorithms and Source Code in C*, John Wiley and Sons, Hoboken, New Jersey, 1996.

[61] M. Stevens, E. Bursztein, P. Karpman, A. Albertini and Y. Markov, The first collision for full SHA-1, *Proceedings of the Thirty-Seventh Annual International Cryptology Conference*, pp. 570–596, 2017.

[62] A. Tridgell, spamsum (www.samba.org/ftp/unpacked/junkcode/spamsum/README), July 27, 2015.

[63] R. van Baar, H. van Beek and E. van Eijk, Digital forensics as a service: A game changer, *Digital Investigation*, vol. 11(S1), pp. S54–S62, 2014.

[64] H. van Beek, E. van Eijk, R. van Baar, M. Ugen, J. Bodde and A. Siemelink, Digital forensics as a service: Game on, *Digital Investigation*, vol. 15, pp. 20–38, 2015.

[65] C. Veenman, Statistical disk cluster classification for file carving, *Proceedings of the Third International Symposium on Information Assurance and Security*, pp. 393–398, 2007.

[66] J. Young, K. Foster, S. Garfinkel and K. Fairbanks, Distinct sector hashes for target file detection, *IEEE Computer*, vol. 45(12), pp. 28–35, 2012.

Chapter 9

ANALYZING WINDOWS SUBSYSTEM FOR LINUX METADATA TO DETECT TIMESTAMP FORGERY

Bhupendra Singh and Gaurav Gupta

Abstract   Timestamp patterns assist forensic analysts in detecting user activities, especially operations performed on files and folders. However, the Windows Subsystem for Linux feature in Windows 10 versions 1607 and later enables users to access and manipulate NTFS files using Linux command-line tools within the Bash shell. Therefore, forensic analysts should consider the timestamp patterns generated by file operations performed using Windows command-line utilities and Linux tools within the Bash shell.

This chapter describes the identification of timestamp patterns of various file operations in stand-alone NTFS and Ext4 filesystems as well as file interactions between the filesystems. Experiments are performed to analyze the anti-forensic capabilities of file timestamp changing utilities – called timestomping tools – on NTFS and Ext4 filesystems. The forensic implications of timestamp patterns and timestomping are also discussed.

Keywords: Anti-forensics, Windows Subsystem for Linux, timestamps, forgery

1.   Introduction

Anti-forensic techniques and tools are increasingly used to circumvent digital forensic investigations. Several definitions of anti-forensics have been proposed over the years [2, 11, 12, 17]. According to Garfinkel [11], anti-forensics seeks to frustrate forensic tools, investigations and investigators. Conlan et al. [6] identify data hiding, encryption, data destruction, steganography and trail obfuscation as notable anti-forensic techniques. In their anti-forensics taxonomy, filesystem manipulation is a subcategory of data hiding. The modification of file timestamps

© IFIP International Federation for Information Processing 2019
Published by Springer Nature Switzerland AG 2019
G. Peterson and S. Shenoi (Eds.): Advances in Digital Forensics XV, IFIP AICT 569, pp. 159–182, 2019.
https://doi.org/10.1007/978-3-030-28752-8_9

– often called "timestomping" – was pioneered by the `timestomp` utility [10]. Timestomping is the intentional alteration of created, modified or accessed timestamps of files or directories in the filesystem of a hard drive, USB stick, flash memory card or other storage device. Because timestamps are vital to event reconstruction and timeline creation, the authenticity and reliability of timestamps extracted from storage media are vital to forensic investigations [3].

The widespread use of anti-forensic tools, such as privacy cleaners (e.g., `CCleaner`) and timestomping utilities (e.g., `SetMACE`), has made the identification and analysis of suspicious files increasingly difficult. Moreover, advanced malware programs use anti-forensic techniques to persist and remain hidden in target systems [19]. These techniques include timestomping, data hiding and filesystem tunneling [5]. For example, Albano et al. [1] have presented an anti-forensic approach that leverages the Linux `touch` command to manipulate the last modified timestamp of an `mmssms.db` database in order to modify or delete evidence in an Android device.

Filesystem metadata analysis is an important component of digital forensic investigations. This chapter focuses on filesystem timestamp patterns that can be used to detect date and time forgery in stand-alone NTFS and Ext4 filesystems as well as in file interactions between the filesystems. The Windows Subsystem for Linux (WSL) feature in Windows 10 enables users to modify, access and delete files and folders in NTFS using Linux tools within a Windows Subsystem for Linux Bash shell. Moreover, users can launch Windows programs and store apps within the shell without leaving (or limiting) evidence in Prefetch files and other sources of program execution data; these traces would have been recorded if similar actions were performed using Windows Explorer. Malicious users can leverage the Windows Subsystem for Linux to manipulate file data and metadata using Linux commands such as `touch` and `shred`. Therefore, it is important to investigate the forensic implications of the Windows Subsystem for Linux feature with a focus on file metadata in hybrid filesystems. The research literature has just two works that discuss the forensic implications of the Windows Subsystem for Linux, one is a blog post [13] and the other is a research article [16].

2.   Filesystem Timestamps

This section presents details about timestamps in the widely-used NTFS and Ext4 filesystems and their time resolutions. The time resolutions of other filesystems, including the newer APFS (Apple Filesystem), are provided to understand anti-forensic techniques that may be used against them.

2.1　NTFS Timestamps and Time Resolutions

NTFS is a complex and robust filesystem used by default in computers running Windows NT 3.1 and later versions. NTFS timestamps are stored as 8-byte file time values that represent the number of 100-nanosecond intervals elapsed since 12:00 A.M. January 1, 1601. Consequently, NTFS timestamps have 100 nanosecond precision.

In addition to MAC (modified, accessed and changed) timestamps, NTFS, unlike other filesystems, also has a birth (i.e., file creation) timestamp. Thus, NTFS metadata contains four types of timestamps for each file on disk: (i) modified (file last modified); (ii) accessed (file last accessed); (iii) created (file created); and (iv) master file table (MFT) entry last modified. These timestamps are commonly referred to by their acronyms – MACE.

Several operating systems allow updates of the access time to be disabled. This means that the access time in a filesystem entry is not updated when the associated file is accessed. However, a user can change the default access time update. For example, the access time in Windows is controlled by the HKLM\SYSTEM\CurrentControlSet\Control \FileSystem\NtfsDisableLastAccessUpdate registry key, where a value of one disables access time updates.

NTFS has two locations in $MFT where timestamps are recorded – $STANDARD_INFORMATION (or $SI) and $FILE_NAME (or $FN). The $SI timestamps are collected by Windows Explorer and by fls, mactime, timestomp, find and other utilities that can display timestamps. The timestamp values in the $SI attribute can be modified by user processes.

NTFS maintains a duplicate set of timestamp values in the $FN attribute. For a given $MFT entry, multiple $FN attributes may exist. Because the $FN attribute can only be modified by the Windows kernel, the MACE values in $FN are updated in a different manner and are inconsistent with the MACE values in the $SI attribute. In fact, timestamp values in $FN are untouched by most timestamp manipulation tools.

It is important to note that the timestamp values in $SI and $FN update independently. For example, when a user accesses or modifies file content, then the temporal values in $SI are updated whereas the temporal values in $FN are updated when the user performs a move or copy operation. Analysis of the temporal values of these two attributes can enable a forensic analyst to detect anomalies and timestomping. Additionally, the information can help an analyst create an accurate timeline of user activities in the system being investigated.

2.2   Ext4 Timestamps and Time Resolutions

Ext4, the successor to the standard Linux filesystem Ext3, is the default filesystem in most Linux distributions. The filesystem has introduced many new features, including the maximum file size (16 GB to 16 TB vs. 16 GB to 2 TB in Ext3), directory capacity (64,000 subdirectories vs. 32,000 in Ext3), journal checksum, journaling feature disabling option, delayed and multiple block allocations, large inodes, used inode count and fast e2fsck and fast extended attributes [7–9, 15]. These features improve the performance and reliability of Ext4 compared with the Ext3 filesystem.

In addition to these features, the inode structure in Ext4 is extended to return seconds and nanoseconds since the Unix epoch (1970-01-01 00:00:00 UTC). Five timestamps are stored in an Ext4 filesystem: (i) file last modification (m-time); (ii) file last access (a-time); (iii) inode metadata changed (c-time); (iv) file creation (cr-time); and (v) file deletion (d-time). The first four timestamps are commonly referred to MACB, where B denotes the birth (or creation) timestamp.

The larger inode structure size of 256 bytes in Ext4 provides additional space to support nanosecond timestamps and postpones the "year 2038 problem" to 2446-05-10 [20]. Mathur et al. [15] have shown that the 32 bits for the c-time, a-time, m-time and cr-time timestamps in Ext3 are extended to 64 bits in Ext4. However, d-time is not extended and remains a 32-bit timestamp in Ext4 with a precision of one second.

The Linux stat API enables users to access MAC timestamps up to nanosecond precision. The regular stat command does not report the cr-time and d-time timestamps. However, users can view all five timestamps with nanosecond precision using the TSK istat command or the debugfs version of stat. Note that the TSK istat command does not consider the extra epoch bits and, therefore, cannot resolve timestamps beyond the year 2038.

Table 1 shows the timestamp resolutions in various filesystems, along with their epoch dates and times.

3.   Experiments and Results

The Windows 10 Anniversary Update (v1607) shipped with the beta version of the Windows Subsystem for Linux feature brings Windows and Linux platforms together. The feature was introduced to reduce

Table 1. Filesystem time resolutions and epoch dates and times.

Filesystem	Timestamps	Size (Bytes)	Resolution	Epoch Date and Time
Ext3	File modified	4	1 s	1970-01-01
	File accessed	4	1 s	00:00:00
	Inode metadata modified	4	1 s	
	File deleted	4	1 s	
FAT32	File modified	4	2 s	1980-01-01
	File accessed	2	1 day	00:00:00
	File created	4	2 s	
Ext4	File modified	8	1 ns	1970-01-01
	File accessed	8	1 ns	00:00:00
	Inode metadata modified	8	1 ns	
	File created	8	1 ns	
	File deleted	4	1 s	
NTFS	File modified	8	100 ns	1601-01-01
	File accessed	8	100 ns	00:00:00
	MFT entry modified	8	100 ns	
	File created	8	100 ns	
HFS+	File modified	4	1 s	1904-01-01
	File accessed	4	1 s	00:00:00 and
	Inode metadata modified	4	1 s	1970-01-01 00:00:00 since
	File created	4	1 s	Mac OS-X 10.7 (Lion)
APFS	File modified	8	1 ns	1970-01-01
	File accessed	8	1 ns	00:00:00
	Inode metadata modified	8	1 ns	
	File created	8	1 ns	

the "gaps" experienced when running Windows tools alongside Linux command-line tools and environments. Since its introduction, the Windows Subsystem for Linux has continuously improved Windows-Linux integration. The Windows 10 Creators Update (v1703) enables users to invoke Windows applications, store apps within a Linux Bash shell and use Linux mainstream developers tools within Windows. Microsoft has stated that it does not explicitly support X/GUI apps/desktops in the

Windows Subsystem for Linux because the intent is only to provide the needed command-line developers tools.

The Windows Subsystem for Linux feature in Windows 10 introduces exciting possibilities for digital forensics. In its Windows 10 Fall Creators Update (v1709), Microsoft announced that the Windows Subsystem for Linux would be a fully-supported operating system feature that would enable users to install multiple Linux distributions and to run them side-by-side simultaneously. Support for mounting USB storage devices is also provided to enable users to access files and folders from within the Linux Bash shell. Clearly, a deep understanding of the Windows Subsystem for Linux feature is required in order for a forensic analyst to correctly interpret timestamp values in filesystems of interest.

3.1   Stand-Alone NTFS Filesystems

The Windows Subsystem for Linux feature in Windows 10 (Anniversary Update and later versions) enables users to perform various file operations in NTFS using mainstream developer command-line tools within the Linux Bash shell. Therefore, an attempt was made to identify timestamp patterns in NTFS for various file operations performed using command-line tools within the Windows Subsystem for Linux. Windows command-line tools and Linux command-line tools were employed to manipulate files. Changes in the MACE timestamps in the $SI and $FN attributes of files were analyzed.

The experimental set-up involved the installation and configuration of Ubuntu and Windows Subsystem for Linux on a personal computer running Windows 10 Pro x64 with version 1709 (build 16299.371). The Ubuntu version was 16.04.03 LTS with core command-line tools (ssh, scp, apt, grep, top, awk, etc.) and mainstream developer tools (emacs, vim, nano, gdb, git, etc.). Following the configuration of Ubuntu, TSK (v4.2.0.3) was installed using apt within the Bash shell. FTK Imager was installed on the Windows system to create the dd image of the NTFS volume. The MACE timestamps in the $SI and $FN attributes corresponding to files of interest were extracted using the istat command.

The experiments focused on several file operations – creation, access, modification, renaming, copying, moving (same volume and across volumes), deletion, compression and decompression. Before performing file operations, several reference files were selected to record their MACE timestamps in $SI and $FN attributes using istat on the dd image of the NTFS volume. Note that istat requires the inode number ($MFT entry number for NTFS) of the file; this was obtained by parsing $MFT using Mft2Csv [18]. Following a file operation, the MACE timestamps were collected for the reference files in the newly-created image of the NTFS volume. This process was repeated for every file operation considered in the timestamp pattern evaluations.

- **Timestamp Rules for File Creation:** These timestamp rules were determined by creating several files using `echo`, `copy` and `fsutil` within the Windows command-line, and `touch` within the Linux Bash shell. It was observed that, whenever a file was created, the MACE timestamps in the $SI and $FN attributes corresponded to the date and time that the file was created.

- **Timestamp Rules for File Access:** These timestamp rules were determined by reopening reference files using their default applications (called the standard GUI mechanism), and using the `cat` and `nano` commands within the Bash shell. It was observed that, when a file in an NTFS volume was accessed using the standard mechanism, then, by default, the em-time timestamp in the $SI attribute was updated to the date and time that the file was last accessed. However, no changes to the MACE timestamps in $FN were observed. Also, when files were accessed using `cat` and `nano`, no changes to the MACE timestamps in the $SI and $FN attributes were observed.

- **Timestamp Rules for File Modification:** These timestamp rules were determined by modifying the reference files using the standard GUI mechanism, and using `echo` and `powershell` within the Windows command-line and `nano` within the Bash shell. The `powershell` command-line shell, which was introduced in Windows 7, also provides a way to modify a file. In all the cases, the m-time and em-time timestamps in the $SI attribute were updated to correspond to the date and time when the file was modified. No changes to the MACE timestamps in $FN were observed.

- **Timestamp Rules for File Renaming:** These timestamp rules were determined by renaming the reference files via Windows Explorer (standard mechanism), `rename` within the Windows command line, and `rename` and `mv` within the Bash shell. It was observed that, in all four cases, em-time in $SI was updated to the date and time when the file was renamed. Moreover, em-time in $FN was changed to the last em-time in $SI. Thus, during normal NTFS operations, em-time in $SI is greater than or equal to em-time in $FN.

- **Timestamp Rules for File Copying:** These timestamp rules were determined by copying and pasting reference files via Windows Explorer, and using `copy` within the Windows command-line and `cp` within the Bash shell. It was observed that, when a copy operation was performed using Windows Explorer or `copy`, the m-time and em-time timestamps in $SI were inherited from the original file. However, the a-time and c-time timestamps were changed to the date and time when the copied file was created on disk. Also, all four timestamps in $FN were updated in every case. When the `cp` command was used within the Bash shell, all the MACE timestamps in $SI and $FN attributes were changed.

- **Timestamp Rules for File Moving:** These timestamp rules were determined by moving reference files within the same NTFS volume and to a different NTFS volume using Windows Explorer, `move` within the Windows command-line and `mv` within the Bash shell. It was discovered that, when a file was moved using the standard mechanism within the same volume, only the $SI em-time was updated to the date and time when the file was moved. No changes were observed to the MACE timestamps in $FN when files were moved to the same volume. However, when a file was moved using `move` or `mv`, in addition to the $SI em-time being updated, the $FN em-time was updated to the last $SI em-time.

  In a second experiment, files were moved to a different NTFS volume using the same methods. It was discovered that the Windows `move` command and Bash shell `mv` command produced different timestamp patterns for a given file. Specifically, the Windows `move` preserved a-time and c-time in $SI whereas the Bash shell `mv` preserved only a-time. However, moving files across volumes changed all the MACE timestamps in the $FN attribute.

- **Timestamp Rules for File Deletion:** These rules were determined by deleting reference files using the Windows Explorer `SHIFT+DELETE`, Windows command-line `del` and Bash shell `rm`. It was discovered that, when a file in an NTFS volume was deleted, none of MACE timestamps in $SI and $FN were changed. In other words, it is very difficult to estimate the deletion dates and times of NTFS files from metadata alone.

- **Timestamp Rules for File Compression:** These rules were determined by compressing files and folders in several ways, such as using `WinZip`, a customized `VBScript`, `7-Zip` and Bash shell `tar`. It was discovered that all the file compression methods created a new file that recorded the MACE timestamps in $SI and $FN as the file compression date and time.

- **Timestamp Rules for File Decompression:** These rules were determined by decompressing files using `WinZip`, `7-Zip` and Bash shell `unzip`. It was discovered that different tools yielded different timestamp patterns. For example, when a compressed file was decompressed using `WinZip`, only the $SI em-time was changed to the file decompression date and time whereas the $SI c-time, m-time and a-time were unchanged from the c-time, m-time and a-time before compression. When a compressed file was decompressed using `7-Zip`, the $SI c-time, em-time and a-time were changed to the file decompression date and time whereas m-time was unchanged from the m-time before compression. If the compressed file was decompressed using the Bash shell `unzip`, then the $SI c-time and em-time were changed to the file decompression date and time whereas the m-time and a-time were unchanged from the m-time and a-time before compression, respectively. However, for all the decompression methods used in the experiments, the $FN MACE timestamps were changed to the file decompression date and time.

Tables 2 through 4 summarize the patterns observed for various operations on NTFS files.

3.2   Stand-Alone Ext4 Filesystems

Several experiments were performed to determine timestamp patterns for various file operations in an Ext4 filesystem.

- **Timestamp Rules for File Creation:** These rules were determined by creating several files using `touch`, `echo` and `cat` within a Ubuntu terminal. It was discovered that, when files were created, the inode MACB timestamps corresponded to the file creation dates and times.

- **Timestamp Rules for File Access:** These rules were determined by reopening reference files using the standard GUI mechanism, and the `cat` and `nano` commands within a Ubuntu terminal. No changes were observed to any of the timestamps.

- **Timestamp Rules for File Modification:** These rules were determined by modifying reference files using the standard GUI mechanism, and the `nano` and `vim` commands within a Ubuntu terminal. It was observed that, in all cases, m-time, a-time and c-time were updated to the dates and times when the files were modified. However, cr-time was unchanged during file modification.

Table 2. Timestamp patterns observed for operations on files in NTFS.

Operation	Method	NTFS Location	Timestamps			
			File Modified (m-time)	File Accessed (a-time)	Entry Modified (em-time)	File Created (c-time)
Creation	Standard GUI	$SI	Creation date, time	Creation date, time	Creation date, time	Creation date, time
		$FN	Creation date, time	Creation date, time	Creation date, time	Creation date, time
	echo command-line	$SI	Creation date, time	Creation date, time	Creation date, time	Creation date, time
		$FN	Creation date, time	Creation date, time	Creation date, time	Creation date, time
	copy command-line	$SI	Creation date, time	Creation date, time	Creation date, time	Creation date, time
		$FN	Creation date, time	Creation date, time	Creation date, time	Creation date, time
	fsutil command-line	$SI	Creation date, time	Creation date, time	Creation date, time	Creation date, time
		$FN	Creation date, time	Creation date, time	Creation date, time	Creation date, time
	touch Bash shell	$SI	Creation date, time	Creation date, time	Creation date, time	Creation date, time
		$FN	Creation date, time	Creation date, time	Creation date, time	Creation date, time
Access	Standard GUI	$SI	Not changed	Not changed	Access date, time	Not changed
		$FN	Not changed	Not changed	Not changed	Not changed
	cat Bash shell	$SI	Not changed	Not changed	Not changed	Not changed
		$FN	Not changed	Not changed	Not changed	Not changed
	nano Bash shell	$SI	Not changed	Not changed	Not changed	Not changed
		$FN	Not changed	Not changed	Not changed	Not changed
Modification	Standard GUI	$SI	Modification date, time	Not changed	Modification date, time	Not changed
		$FN	Not changed	Not changed	Not changed	Not changed
	echo command-line	$SI	Modification date, time	Not changed	Modification date, time	Not changed
		$FN	Not changed	Not changed	Not changed	Not changed
	powershell command-line	$SI	Modification date, time	Not changed	Modification date, time	Not changed
		$FN	Not changed	Not changed	Not changed	Not changed
	nano Bash shell	$SI	Modification date & time	Not changed	Modification date & time	Not changed
		$FN	Not changed	Not changed	Not changed	Not changed

Table 3. Timestamp patterns observed for operations on files in NTFS (continued).

Operation	Method	NTFS Location	Timestamps			
			File Modified (m-time)	File Accessed (a-time)	Entry Modified (em-time)	File Created (c-time)
Rename	Standard	$SI	Not changed	Not changed	Rename date, time	Not changed
	Windows Explorer	$FN	Not changed	Not changed	Not changed	Not changed
	rename	$SI	Not changed	Not changed	Rename date, time	Not changed
	command-line	$FN	Not changed	Not changed	Last $SI em-time	Not changed
	rename	$SI	Not changed	Not changed	Rename date, time	Not changed
	Bash shell	$FN	Not changed	Not changed	Last $SI em-time	Not changed
	mv	$SI	Not changed	Not changed	Rename date, time	Not changed
	Bash shell	$FN	Not changed	Not changed	Last $SI em-time	Not changed
Copy	Standard	$SI	Not changed	Copy date, time	Not changed	Copy date, time
	GUI	$FN	Copy date, time	Copy date, time	Copy date, time	Copy date, time
	copy	$SI	Not changed	Copy date, time	Not changed	Copy date, time
	command-line	$FN	Copy date, time	Copy date, time	Copy date, time	Copy date, time
	cp	$SI	Copy date, time	Copy date, time	Copy date, time	Copy date, time
	Bash shell	$FN	Copy date, time	Copy date, time	Copy date, time	Copy date, time
Move (Same Volume)	Standard	$SI	Not changed	Not changed	Move date, time	Not changed
	Windows Explorer	$FN	Not changed	Not changed	Last $SI em-time	Not changed
	move	$SI	Not changed	Not changed	Move date, time	Not changed
	command-line	$FN	Not changed	Not changed	Last $SI em-time	Not changed
	mv	$SI	Not changed	Not changed	Move date, time	Not changed
	Bash shell	$FN	Not changed	Not changed	Last $SI em-time	Not changed
Move (Across Volumes)	Standard	$SI	Not changed	Move date, time	Not changed	Move date, time
	Windows Explorer	$FN	Move date & time	Move date, time	Move date & time	Move date, time
	move	$SI	Not changed	Move date, time	Not changed	Move date, time
	command-line	$FN	Move date, time	Move date, time	Move date, time	Move date, time
	mv	$SI	Move date & time	Not changed	Move date & time	Move date, time
	Bash shell	$FN	Move date, time	Move date, time	Move date, time	Move date, time

Table 4. Timestamp patterns observed for operations on files in NTFS (continued).

Operation	Method	NTFS Location	Timestamps			
			File Modified (m-time)	File Accessed (a-time)	Entry Modified (em-time)	File Created (c-time)
Deletion	Standard Windows Explorer	$SI	Not changed	Not changed	Not changed	Not changed
		$FN	Not changed	Not changed	Not changed	Not changed
	SHIFT+DELETE	$SI	Not changed	Not changed	Not changed	Not changed
		$FN	Not changed	Not changed	Not changed	Not changed
	del command-line	$SI	Not changed	Not changed	Not changed	Not changed
		$FN	Not changed	Not changed	Not changed	Not changed
	rm Bash shell	$SI	Not changed	Not changed	Not changed	Not changed
		$FN	Not changed	Not changed	Not changed	Not changed
Compression	WinZip and 7-Zip	$SI	Comp. date, time	Comp. date, time	Comp. date, time	Comp. date, time
		$FN	Comp. date, time	Comp. date, time	Comp. date, time	Comp. date, time
	VBScript command-line	$SI	Comp. date, time	Comp. date, time	Comp. date, time	Comp. date, time
		$FN	Comp. date, time	Comp. date, time	Comp. date, time	Comp. date, time
	tar Bash shell	$SI	Comp. date, time	Comp. date, time	Comp. date, time	Comp. date, time
		$FN	Comp. date, time	Comp. date, time	Comp. date, time	Comp. date, time
Decompression	WinZip	$SI	File last m-time	File last a-time	Decomp. date, time	File last c-time
		$FN	Decomp. date, time	Decomp. date, time	Decomp. date, time	Decomp. date, time
	7-Zip	$SI	File last m-time	Decomp. date, time	Decomp. date, time	Decomp. date, time
		$FN	Decomp. date, time	Decomp. date, time	Decomp. date, time	Decomp. date, time
	unzip	$SI	File last m-time	File last a-time	Decomp. date, time	Decomp. date, time
	Bash shell	$FN	Decomp. date, time	Decomp. date, time	Decomp. date, time	Decomp. date, time

- **Timestamp Rules for File Renaming:** These rules were determined by renaming reference files using the standard GUI mechanism, and the `rename` and `mv` commands within a Ubuntu terminal. It was discovered that a-time and c-time were updated to the date and time when the renaming operation was performed whereas m-time and cr-time were unchanged.

- **Timestamp Rules for File Copying:** These rules were determined by copying reference files using the standard GUI mechanism and the `cp` command within a Ubuntu terminal. It was discovered that the m-time of a copied file was inherited from the original file whereas the a-time, c-time and cr-time were updated to the date and time of the copy operation. However, if the file was copied using the `cp` command, then all the MACB timestamps were changed to the date and time of the copy operation.

- **Timestamp Rules for File Moving:** These rules were determined by moving reference files to the same Ext4 volume as well as to another Ext4 volume using the standard GUI mechanism and the `mv` command within a Ubuntu terminal. In all cases, a-time and c-time were changed to the date and time of the move operation. However, m-time and cr-time were unchanged after all the move operations.

- **Timestamp Rules for File Deletion:** These rules were determined by deleting files using the standard `SHIFT+DELETE`, and the `rm` and `shred` commands within a Ubuntu terminal. It was discovered that, in addition to the m-time and c-time, the Ext4 filesystem recorded the d-time of the particular file, and all the timestamp values were updated to the date and time when the file was deleted. However, a-time and cr-time were unchanged after all the deletion operations.

- **Timestamp Rules for File Compression:** These rules were determined by compressing files and folders in an Ext4 volume using the standard GUI mechanism, and the `zip`, `tar` and `gzip` commands within a Linux terminal. It was discovered that, if a file or folder was compressed using the standard GUI mechanism or using `zip` or `tar`, then the inode MACB timestamps stored in the volume corresponded to the date and time when the file or folder was compressed. However, different timestamp patterns were observed for different compression algorithms. For example, when the `gzip` command-line tool within a Ubuntu Bash terminal was used, m-time and a-time were not updated. However, c-time and cr-time were updated to the date and time when the compression command was executed.

- **Timestamp Rules for File Decompression:** These rules were determined by decompressing several files using the standard GUI mechanism, and the unzip, tar and gzip commands within a Ubuntu terminal. It was discovered that, when a compressed file was extracted or decompressed, the timestamp patterns depended on the decompression method used. For example, unzip and tar left m-time unchanged from the file last modified time (just before compression); the other three timestamps were updated to the date and time when the zipped file was decompressed. In the case of gzip, m-time and a-time were unchanged.

Tables 5 and 6 summarize the patterns observed for various operations on Ext4 files.

3.3   NTFS-Ext4 File Transfers

Experiments were conducted to identify the timestamp patterns of file transfers to and from Windows NTFS and Linux Ext4 volumes. In the experiments, a personal computer was set up to dual boot with Microsoft Windows 10 v1709 x64 and Ubuntu 16.04.03 LTS. The TSK tool was installed using apt within a Ubuntu Bash terminal. To enable file transfers between NTFS and Ext4, the Windows volumes were mounted in Ubuntu and files were transferred using methods such as the standard GUI mechanism, and cp and mv within a Ubuntu Bash terminal. The istat tool was used to collect timestamps before and after each file was transferred.

In the case of file transfers from NTFS to Ext4 using the standard GUI, and cp and mv within a Ubuntu Bash terminal, it was discovered that, for all the cases shown in Table 7, at least a-time, c-time and cr-time were changed to the file transfer date and time. However, file transfers using cp changed all the inode MACB timestamps. Also, file transfers from NTFS to Ext4 using the standard GUI and mv within a Ubuntu Bash terminal caused the inode m-time to be inherited from $SI m-time.

When a file was transferred from Ext4 to NTFS using the standard GUI, and cp and mv within a Ubuntu Bash terminal, all the MACE timestamps in $FN were assigned the date and time when the file was transferred. Also, a-time, em-time and c-time in $SI were updated to the file transfer date and time. However, except for the file transfer using cp, m-time in $SI was not updated. In fact, it inherited the inode m-time of

Table 5. Timestamp patterns observed for operations on files in Ext4.

Operation	Method	Timestamps				
		File Modified (m-time)	File Accessed (a-time)	Inode Modified (c-time)	File Created (cr-time)	File Deleted (d-time)
Creation	touch	Creation date, time	Creation date, time	Creation date, time	Creation date, time	NA
	cat	Creation date, time	Creation date, time	Creation date, time	Creation date, time	NA
	echo	Creation date, time	Creation date, time	Creation date, time	Creation date, time	NA
Access	GUI	Not changed	Not changed	Not changed	Not changed	NA
	cat	Not changed	Not changed	Not changed	Not changed	NA
	nano	Not changed	Not changed	Not changed	Not changed	NA
Modification	GUI	Modification date, time	Modification date, time	Modification date, time	Not changed	NA
	nano	Modification date, time	Modification date, time	Modification date, time	Not changed	NA
	vim	Modification date, time	Modification date, time	Modification date, time	Not changed	NA
Rename	GUI	Not changed	Rename date, time	Rename date, time	Not changed	NA
	rename	Not changed	Rename date, time	Rename date, time	Not changed	NA
	mv	Not changed	Rename date, time	Rename date, time	Not changed	NA
Copy	GUI	Not changed	Copy date, time	Copy date, time	Copy date, time	NA
	cp	Copy date & time	Copy date, time	Copy date, time	Copy date, time	NA
Move	GUI	Not changed	Move date, time	Move date, time	Not changed	NA
	Move To Trash	Not changed	Move date, time	Move date, time	Not changed	NA
	mv	Not changed	Move date, time	Move date, time	Not changed	NA

Table 6. Timestamp patterns observed for operations on files in Ext4.

Operation	Method	Timestamps				
		File Modified (m-time)	File Accessed (a-time)	Inode Modified (c-time)	File Created (cr-time)	File Deleted (d-time)
Deletion	Move to Trash and Delete	Del. date, time	Not changed	Del. date, time	Not changed	Del. date, time
	SHIFT+DELETE	Del. date, time	Not changed	Del. date, time	Not changed	Del. date, time
	rm	Del. date, time	Not changed	Del. date, time	Not changed	Del. date, time
	shred	Del. date, time	Not changed	Del. date, time	Not changed	Del. date, time
Compression	GUI	Comp. date, time	Comp. date, time	Comp. date, time	Comp. date, time	NA
	tar	Comp. date, time	Comp. date, time	Comp. date, time	Comp. date, time	NA
	zip	Comp. date, time	Comp. date, time	Comp. date, time	Comp. date, time	NA
	gzip	Not changed	Not changed	Comp. date, time	Comp. date, time	NA
Decompression	GUI	File last m-time	Decomp. date, time	Decomp. date, time	Decomp. date, time	NA
	unzip	File last m-time	Decomp. date, time	Decomp. date, time	Decomp. date, time	NA
	tar	File last m-time	Decomp. date, time	Decomp. date, time	Decomp. date, time	NA
	gzip	Not changed	Not changed	Decomp. date, time	Decomp. date, time	NA

Table 7. Timestamp patterns observed during NTFS to Ext4 file transfers.

Operation	Transfer Method	Timestamps			
		File Modified (m-time)	File Accessed (a-time)	Inode Modified (c-time)	File Created (cr-time)
Copy	File copy via GUI	Not changed	Copy date and time	Copy date and time	Copy date and time
Copy	File copy via cp in Bash shell	Copy date and time	Copy date and time	Copy date and time	Copy date and time
Move	File move via GUI	Not changed	Move date and time	Move date and time	Move date and time
Move	File move via mv in Bash shell	Not changed	Move date and time	Move date and time	Move date and time

the file in the Ext4 volume. Table 8 summarizes the timestamp patterns observed during the Ext4 to NTFS file transfers.

3.4   Timestomping Tool Capabilities

Experiments were conducted with several timestomping utilities and command-line tools to imvestigate evidence tampering in NTFS and Ext4 filesystems.

Six utilities, `BulkFileChanger`, `Attribute Changer`, `SKTimeStamp`, `FS Touch`, `SetMACE` and `AttributeMagic`, were used to alter the MACE timestamps in an NTFS volume. The utilities were installed on a Windows 10 system and various (newly created and existing) files in a disk volume were considered as reference files.

Before any timestamps were modified, the MACE timestamps corresponding to $SI and $FN were collected from the reference files in an NTFS volume image (i.e., `dd` image created using FTK Imager). The timestamps were extracted using the TSK `istat` command within a Bash shell. The inode number (MFT entry number for NTFS) of a file required by `istat` was obtained by parsing $MFT using `Mft2Csv`.

Each of the six anti-forensic utilities was executed to change the timestamps of the reference files to future times. A `dd` image of the NTFS volume was then created and `istat` was executed to extract MACE timestamps from the $SI and $FN attributes of the reference files. It

Table 8. Timestamp patterns observed during Ext4 to NTFS file transfers.

Operation	Transfer Method	NTFS Location	Timestamps			
			File Modified (m-time)	File Accessed (a-time)	Entry Modified (em-time)	File Created (c-time)
Copy	File copy via GUI	$SI	Not changed	Copy date and time	Copy date and time	Copy date and time
		$FN	Copy date and time	Copy date and time	Copy date and time	Copy date and time
Copy	File copy via cp in Bash shell	$SI	Copy date and time	Copy date and time	Copy date and time	Copy date and time
		$FN	Copy date and time	Copy date and time	Copy date and time	Copy date and time
Move	File move via GUI	$SI	Not changed	Move date and time	Move date and time	Move date and time
		$FN	Move date and time	Move date and time	Move date and time	Copy date and time
Move	File move via cp in Bash shell	$SI	Note changed	Move date and time	Move date and time	Move date and time
		$FN	Move date and time	Move date and time	Move date and time	Move date and time

is believed that is the best methodology for comparing the anti-forensic capabilities of timestomping utilities.

By and large, the six timestomping tools could not set the $FN MACE timestamps. In fact, only the SetMACE command-line tool could alter the $FN MACE timestamps. Additionally, whereas all the tools could alter the $SI MAC timestamps, no tool – except for SetMACE – could alter the $SI em-time. However, BulkFileChanger, Attribute Changer, SKTimeStamp, FS Touch and AttributeMagic set the em-time to the date and time when the tool was executed. These five tools were unable to set the nanoseconds portion of the date and time; instead, they set all nine digits after the seconds part to zeroes. FS Touch could only change the $SI MAC timestamps up to the milliseconds part (three digits after the seconds part).

Only the SetMACE command-line tool was able to successfully alter all the MACE timestamps in $SI and $FN with nanosecond precision. This is because, unlike the other tools, SetMACE performs direct disk accesses to manipulate the MACE timestamps in $SI and $FN, as well as in the $INDEX_ROOT and $INDEX_ALLOCATION attributes. SetMACE accomplishes this using a driver that bypasses the filesystem and writes directly

Table 9.   Timestomping capabilities of six anti-forensic tools on NTFS.

Tool	Timestamps	File Modified	File Accessed	Entry Modified	File Created
BulkFileChanger	$SI	✓	✓	●	✓
v1.51	$FN	✗	✗	✗	✗
Attribute	$SI	✓	✓	●	✓
Changer v9.0a	$FN	✗	✗	✗	✗
SKTimeStamp	$SI	✓	✓	●	✓
v1.3.5	$FN	✗	✗	✗	✗
FS Touch v7.3	$SI	✓	✓	●	✓
	$FN	✗	✗	✗	✗
SetMACE	$SI	✔	✔	✔	✔
v1.0.0.14	$FN	✔	✔	✔	✔
AttributeMagic	$SI	✓	✓	●	✓
v2.4	$FN	✗	✗	✗	✗

to the disk without leaving any traces in NTFS metadata ($LogFile and $UsnJrnl), provided that the adversary has elevated disk access privileges. That means that SetMACE resolves the filesystem internally and writes the timestamps directly to the physical disk, bypassing filesystem and operating system control mechanisms. As a result, detecting traces of SetMACE execution is extremely difficult.

Table 9 compares the timestomping capabilities of the six anti-forensic tools on NTFS. The ✓ symbol denotes that the timestamp was changed, but the nanoseconds part was zeroed. The ● symbol indicates that the timestamp was changed to the date and time when the utility was executed. The ✓ symbol denotes that the timestamp was changed, but only up to the milliseconds part. The ✔ symbol signifies that the timestamp was changed, including the nanoseconds part. Finally, the ✗ symbol indicates that the timestamp was not changed.

Five utilities, chmod, chattr, touch, SetMACB and BulkFileChanger, were used to alter the MACB timestamps in an Ext4 filesystem. TSK was installed on a computer running Ubuntu 16.04 LTS and several (newly created and existing) files in the Ext4 volume were considered as reference files. BulkFileChanger was executed on the Ubuntu 16.04 LTS system using the Wine package, which is capable of running Windows applications on several POSIX-compliant operating systems. Before any timestamps were modified, the MACB timestamps corresponding to the reference files in the Ext4 volume were extracted using stat (when the file inode number was known) and `istat` (otherwise). Following this, each tool was executed to manipulate the MACB timestamps of the reference files. The new timestamps then were recorded.

The analysis revealed that `chmod` updated only the accessed (a-time) and inode modified (c-time) timestamps whereas `chattr` only updated c-time. The `touch` command (with default options) updated a-time and c-time to the current date and time. The `touch` command was also able to alter the accessed and file modified timestamps to a specific date and time using the -a option for a-time and the -m option for m-time. For example, `touch -a -m -t 201612061104.45 test.txt` changed the m-time and a-time of `test.txt` to 2016-12-06 11:04:45.000000000, but c-time was updated to the date and time when the command was executed. If the system date and time were correct, then c-time would have the correct date and time; however, if the system time was manipulated, then c-time would have the incorrect date and time.

The `touch` command also changed m-time and a-time up to the nanoseconds part using the -d option. For example, `touch -d "2016-12-06 11:04:45.123456789" test.txt` changed the m-time and a-time of the file `test.txt` to 2016-12-06 11:04:45.123456789, but c-time was updated to the date and time when the command was executed.

The `touch` command also copied the timestamps of a file to a target file using option -r. For example, `touch test.txt -r sample.txt` set the m-time of `test.txt` file to the m-time of `sample.txt`. Also, the a-time and c-time of the `test.txt` file were updated to the date and time when the command was executed; however, the created timestamp (cr-time) was untouched.

The `BulkFileChanger` utility was unable to manipulate cr-time; it only changed the date and time of m-time and left the nanoseconds part to be all zeroes.

However, it is possible to manipulate all the MACB timestamps of a file in an Ext4 filesystem. A workaround procedure, referred to as `SetMACB` in Table 10, was created to successfully manipulate all four MACB timestamps with nanosecond precision. The `SetMACB` procedure involved the following steps:

- Alter the system time to the desired date and time using the `date` command-line tool.

- Create a new file with the same content or copy-paste an existing file whose timestamps need to be manipulated.

- Update the a-time and m-time of the file using the `touch` command-line tool.

Table 10.   Timestomping capabilities of five anti-forensic tools on Ext4 filesystems.

Tool	Options	File Modified	File Accessed	Inode Modified	File Created
chmod	775	✗	✗	●	✗
chattr	+a	✗	✗	●	✗
touch	-a -t	✗	✓	●	✗
	-a -m -t	✓	✓	●	✗
	-d	✔	✔	●	✗
	-r	✔	✔	●	✗
SetMACB	See text	✔	✔	✔	✔
BulkFileChanger	Using Wine	✓	●	●	✗

- Reset the system time to the current date and time using the date command-line tool.

Table 10 compares the timestomping capabilities of the five anti-forensic tools on the Ext4 filesystem. The ✓ symbol denotes that the timestamp was changed, but the nanoseconds part was zeroed. The ● symbol indicates that the timestamp was changed to the date and time when the utility was executed. The ✔ symbol signifies that the timestamp was changed, including the nanoseconds part. Finally, the ✗ symbol indicates that the timestamp was not changed.

4.   Discussion

Analyzing timestamps in a filesystem to their full precision is important to detect timestamp forgery. The experiments reveal that, by and large, the evaluated timestomping tools could set the file created, modified and accessed timestamps to specified dates and times with precisions of seconds, leaving the nanoseconds parts as zeroes. Thus, if the nanoseconds part of any NTFS MACE or Ext4 MACB timestamp (except for inode d-time in Ext4) contains all zeroes, then timestamp forgery is indicated. However, this is not always true because, when a file is copied or moved from a USB device (FAT32 filesystem) to an NTFS or Ext4 volume, it inherits the m-time (in $SI only for NTFS), but the nanoseconds part has all zeroes because the time resolution of the last modified time in a FAT32 filesystem only has a precision of seconds. In the case of an NTFS file, because of the 100 ns interval, there is one chance out of $4,782,969 (= 9^7)$ that the nanoseconds part of a timestamp would be all zeroes by default. In the case of an Ext4 file, the chance is only one out of $387,420,489 (= 9^9)$.

The experiments also reveal that the `touch` timestamp manipulation tool behaves differently in the NTFS and Ext4 filesystems. In the case of NTFS, `touch` (by default), updates a-time, m-time and em-time to the system date and time in the $SI attribute whereas all four timestamps in the $FN attribute are not touched. In the case of Ext4, `touch` updates a-time, m-time and c-time to the system date and time. However, `touch` does not manipulate file creation timestamps in both the filesystems.

5.   Conclusions

Timestamp patterns assist forensic analysts in detecting user activities in filesystems, especially operations performed on files. However, anti-forensic techniques such as timestomping can alter file created, modified and accessed timestamps in the filesystems of hard drives, USB sticks, flash memory cards and other storage devices. Because timestamps are vital to event reconstruction and timeline creation, the determination of the authenticity and reliability of timestamps extracted from storage media are vital in forensic investigations

The filesystem timestamp patterns specified in this chapter enable forensic analysts to detect date and time forgeries in stand-alone NTFS and Ext4 filesystems as well as forgeries related to file transfers between the two filesystems. The analysis of well-known file timestamp changing utilities (timestomping tools) on NTFS and Ext4 filesystems provides valuable insights into their anti-forensic capabilities. Timestamp anomalies can be detected by leveraging timestamp patterns and analyzing timestamps to their full precision.

The research described in this chapter has focused on the Windows Subsystem for Linux feature in Windows 10 systems. Since this feature is still evolving, it is expected that the timestamp patterns would vary in future versions of Windows 10. Nevertheless, this research has demonstrated how timestamp patterns and the capabilities of anti-forensic tools can be systematically investigated to detect timestamp forgeries.

Future research will investigate the forensic implications of the Windows Subsystem for Linux feature with regard to the recovery of deleted files using Linux tools such as `rm`, `shred` and `srm`. Also, research will attempt to identify the sources and locations of execution artifacts created when Windows programs and apps are launched within a Bash shell using the Windows Subsystem for Linux feature.

References

[1] P. Albano, A. Castiglione, G. Cattaneo and A. De Santis, A novel anti-forensic technique for the Android OS, *Proceedings of the International Conference on Broadband and Wireless Computing, Communications and Applications*, pp. 380–385, 2011.

[2] I. Baggili, A. BaAbdallah, D. Al-Safi and A. Marrington, Research trends in digital forensic science: An empirical analysis of published research, *Proceedings of the Fourth International Conference on Digital Forensics and Cyber Crime*, pp. 144–157, 2012.

[3] F. Buchholz and E. Spafford, On the role of filesystem metadata in digital forensics, *Digital Investigation*, vol. 1(4), pp. 298–309, 2004.

[4] B. Carrier, *File System Forensic Analysis*, Pearson Education, Upper Saddle River, New Jersey, 2005.

[5] E. Casey, Digital stratigraphy: Contextual analysis of filesystem traces in forensic science, *Journal of Forensic Sciences*, vol. 63(5), pp. 1383–1391, 2018.

[6] K. Conlan, I. Baggili and F. Breitinger, Anti-forensics: Furthering digital forensic science through a new, extended, granular taxonomy, *Digital Investigation*, vol. 18(S), pp. S66–S75, 2016.

[7] A. Dewald and S. Seufert, AFEIC: Advanced forensic Ext4 inode carving, *Digital Investigation*, vol. 20(S), pp. S83–S91, 2017.

[8] K. Fairbanks, An analysis of Ext4 for digital forensics, *Digital Investigation*, vol. 9(S), pp. S118–S130, 2012.

[9] K. Fairbanks, C. Lee and H. Owen III, Forensic implications of Ext4, *Proceedings of the Sixth Annual Workshop on Cyber Security and Information Intelligence Research*, article no. 22, 2010.

[10] J. Foster and V. Liu, Catch me, if you can, presented at *Black Hat Japan*, 2005.

[11] S. Garfinkel, Anti-forensics: Techniques, detection and countermeasures, *Proceedings of the Second International Conference on i-Warfare and Security*, pp. 77–84, 2007.

[12] R. Harris, Arriving at an anti-forensics consensus: Examining how to define and control the anti-forensics problem, *Digital Investigation*, vol. 3(S), pp. 44–49, 2006.

[13] A. Harrison, Further Forensicating of Windows Subsystem for Linux, *1234n6 Blog* (www.blog.1234n6.com/2017/10/further-forensicating-of-windows.html), October 17, 2017.

[14] S. Ho, D. Kao and W. Wu, Following the breadcrumbs: Timestamp pattern identification for cloud forensics, *Digital Investigation*, vol. 24, pp. 79–94, 2018.

[15] A. Mathur, M. Cao, S. Bhattacharya, A. Dilger, A. Tomas and L. Vivier, The new Ext4 filesystem: Current status and future plans, *Proceedings of the Linux Symposium*, vol. 2, pp. 21–34, 2007.

[16] L. Nathan, A. Case, A. Ali-Gombe and G. Richard III, Memory forensics and the Windows Subsystem for Linux, *Digital Investigation*, vol. 26(S), pp. S3–S11, 2018.

[17] M. Rogers, Anti-forensics: The coming wave in digital forensics, poster presentation at the *Seventh Annual CERIAS Information Security Symposium*, 2006.

[18] J. Schicht, Mft2Csv, GitHub (www.github.com/jschicht/Mft2 Csv/wiki/Mft2Csv), May 20, 2017.

[19] B. Singh and U. Singh, Program execution analysis in Windows: A study of data sources, their formats and comparison of forensic capability, *Computers and Security*, vol. 74, pp. 94–114, 2018.

[20] D. Wong, Ext4 Disk Layout (www.ext4.wiki.kernel.org/index.php/Ext4_Disk_Layout), February 18, 2019.

IV

IMAGE FORENSICS

Chapter 10

QUICK RESPONSE ENCODING OF HUMAN FACIAL IMAGES FOR IDENTITY FRAUD DETECTION

Shweta Singh, Saheb Chhabra, Garima Gupta, Monika Gupta and Gaurav Gupta

Abstract   Advancements in printing and scanning technology enable fraudsters to tamper with identity documents such as identity cards, drivers' licenses, admit cards, examination hall tickets and academic transcripts. Several security features are incorporated in important identity documents to counter forgeries and verify genuineness, but these features are often lost in printed versions of the documents. At this time, a satisfactory method is not available for authenticating a person's facial image (photograph) in a printed version of a document. Typically, an official is required to check the person's image against an image stored in an online verification database, which renders the problem even more challenging.

This chapter presents an automated, low-cost and efficient method for addressing the problem. The method employs printed quick response codes corresponding to low-resolution facial images to authenticate the original and printed versions of identity documents.

Keywords: Facial images, documents, quick response codes, tamper detection

1.   Introduction

Advancements in printing and scanning technology have made it easy for fraudsters to produce high-quality tampered documents. Indeed, changing or replacing the facial image of a person on an identity document is becoming very common.

Numerous frauds have been perpetrated by tampering with human facial images on documents. One example is the Vyapam scam in India [10], where a fraudster replaces the photograph of a student on an

© IFIP International Federation for Information Processing 2019
Published by Springer Nature Switzerland AG 2019
G. Peterson and S. Shenoi (Eds.): Advances in Digital Forensics XV, IFIP AICT 569, pp. 185–199, 2019.
https://doi.org/10.1007/978-3-030-28752-8_10

examination admit card with that of an imposter who takes the exam on behalf of the student. Another example comes from China [15], where a broker finds a proxy to take an important exam such as the SAT, GRE or GMAT on behalf of a student. The broker then prepares a fake passport with the information of the student but with the image of the proxy.

Authenticating a tampered identity document requires an expert to manually analyze the document using sophisticated equipment such as a microscope or video spectral comparator. This process is time-consuming, inefficient and non-scalable. Also, this method for detecting tampered documents is not applicable to printed versions of documents because most security features are lost during the printing process.

Clearly, there is a need to develop an automated system that can authenticate a person's facial image on a document. This system should work for originals as well as printed versions of documents. Also, the system should be able to authenticate documents offline and without relying on a database of images. Additionally, if tampering is detected, the system should be able to reproduce the person's facial image that is similar to the original image in the document.

This chapter presents a method that employs printed quick response (QR) codes of low-resolution facial images to authenticate the original and printed versions of documents. The method supports image-to-image verification, which matches an image on an identity document against the image encoded as a quick response code on the same document. Also, it supports real-time person verification, which matches an image encoded as a quick response code on the identity document against a person's image captured in real time. Figure 1 shows an example of authenticating (and detecting the tampering of) a driver's license using the proposed method.

2.   Related Work

Several researchers have proposed methods for detecting counterfeit documents. Gupta et al. [6] describe a method that considers the texture and unique color count in order to detect counterfeit documents. The method links a counterfeit document to its source scanner and printer. While the method can differentiate between the original and printed versions of documents, it is not effective at detecting tampering in printed documents.

Sarkar et al. [12] have proposed a method for detecting low-quality and high-quality counterfeit currency notes. They have also analyzed printed security features on currency notes.

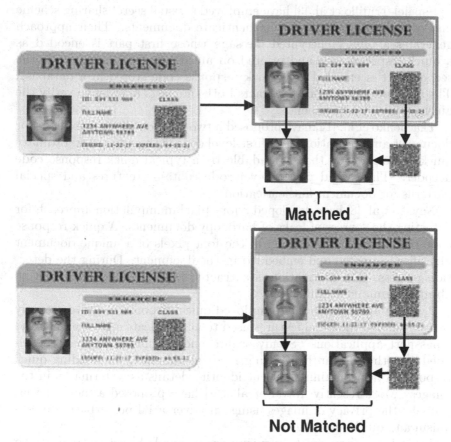

Figure 1. Authentication of a facial image on a driver's license.

Chhabra et al. [2] have developed a method for detecting fraudulent bank checks, including checks whose text has been altered using invisible ink. However, this method for alteration detection does not work well for printed documents.

The method proposed in this chapter employs quick response codes as information carriers. In the research literature, quick response codes have been used in applications ranging from physical document authentication to securing information and identifying leaked documents.

Warasart and Kuacharoen [16] have used quick response codes to authenticate text-based physical documents. Specifically, they created codes based on text in documents and verified them against codes created using text extracted from the documents by an optical character recognition system.

Espejel-Trujillo et al. [3] have employed a visual secret sharing scheme and quick response codes to authenticate documents. Their approach generates a binary encrypted message whose first part is encoded as a quick response code and printed on an identity document while the second part is encoded as a quick response code stored in a database. The authentication process extracts both the parts in order to verify the authenticity of a document.

Tkachenko et al. [14] have proposed a two-level quick response code for document authentication. The first-level code is the same as an ordinary quick response code that is readable by a typical quick response code decoder. The second private-level code contains textures and special patterns for document authentication.

Nayak et al. [8] have developed a font pixel manipulation approach for detecting the sources of leaks of hardcopy documents. A quick response code that encodes information in the font pixels of a unique document identifier is created and embedded in the document. During the detection process, this information is extracted from a leaked document to obtain its unique identifier.

Aygun and Akcay [1] have employed quick response codes to securely transmit biometric facial features used to authenticate e-government and e-passport applications. Seenivasagam and Velumani [13] have developed a method for authenticating medical images by embedding quick response codes containing patient identity details as watermarks in the images. More recently, Raval et al. [11] have proposed a method that protects the privacy of images using an adversarial perturbation mechanism and quick response codes.

Table 1 compares existing quick-response-code-based methods along with the proposed authentication method. The literature review indicates that no automated approach for authenticating a person's facial image in a document has been published previously. The proposed method addresses the deficiency by performing authentications in an offline manner. Low-resolution facial images are stored as quick response codes on documents when creating the documents. During authentication, the low-resolution image stored as a quick response code on a document is extracted and matched against a person's facial image on the document.

Two challenges are encountered when storing and verifying low-resolution images using quick response codes. First, because a quick response code has limited data storage capacity, an image is downsampled to a low resolution (e.g., 16×16 or 8×8), which leads to considerable information loss. Second, authentication requires the comparison of a low-resolution image against a high-resolution image. To address these challenges, the

Table 1. Comparison of tampering detection methods.

Authors	Method	Documents	Facial Image Authentication
Gupta et al., 2007 [6]	Texture and unique color count	Original	No
Chhabra et al., 2017 [2]	Texture	Original	No
Nayak et al., 2018 [8]	Font pixel manipulation	Printed	No
Espejel-Trujillo et al., 2016 [3]	Visual secret sharing	Printed	No
Singh et al., 2019 (proposed method)	Generative adversarial net	Original and printed	Yes

proposed method uses deep learning to enhance the low-resolution image that is used for authentication.

3. Proposed Method

The proposed method has two steps: (i) document generation; and (ii) document authentication.

3.1 Document Generation

Given a document with a low-resolution facial image of a person, a quick response code corresponding to the image is created. This quick response code is then printed on the document. Figure 2(a) shows the steps involved in document generation.

3.2 Document Authentication

Two common situations are encountered when attempting to authenticate a person's facial image on a document: (i) image-to-image verification; and (ii) real-time person verification.

In image-to-image verification, the low-resolution image already encoded as a quick response code on a document is matched against the person's image on the document. This is required when the person's facial image on the document must be verified for possible tampering.

In real-time person verification, the facial image already encoded as a quick response code on a document is matched against the person's facial image captured in real time. This is required when it is difficult to verify

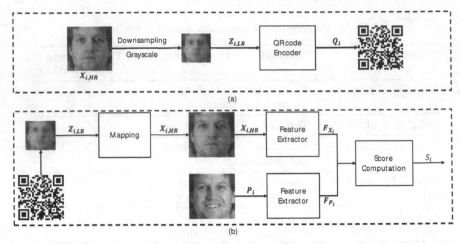

Figure 2.   Proposed method.

the facial image on the document as a result of changes in the person's facial features due to age or injury or when the person's image on the document has been damaged. Figure 2(b) shows the steps involved in document authentication using the quick response code computed from a low-resolution image.

The embedded quick response code and the person's image on the document make it possible to perform the authentication process offline (i.e., without a database). The research literature demonstrates that quick response codes work well on printed documents.

The document authentication step has two components: (i) low-resolution to high-resolution image mapping; and (ii) image matching:

- **Low-Resolution to High-Resolution Image Mapping:** A generative adversarial network model [4] is employed to transform a low-resolution grayscale image encoded as a quick response code to a high-resolution color image. The approach involves training a generator model G to produce an output image x from a noise vector $\mathbf{z}$, and a discriminator D to distinguish between the real image y and a generated image x. The objective of the generator is to produce images such that the discriminator cannot distinguish between the real images and the generated images, where the discriminator has been trained to distinguish real images from generated images.

  The objective function of a generative adversarial network is given by:

$$\mathcal{L}_{GAN}(G, D) = \mathbb{E}_y[logD(y)] + \mathbb{E}_z[log(1 - D(G(z)))] \quad (1)$$

where the generator model G tries to minimize the objective function and the discriminator D tries to maximize the objective function.

A conditional generative adversarial network is trained to transform the input low-resolution grayscale image to a high-resolution color image.

Let $\mathbf{X_{HR}}$ be a training set containing m high-resolution color images associated with a document. Let $\mathbf{Z_{LR}}$ be the corresponding low-resolution grayscale image set. Each image $\mathbf{Z_{i,LR}}$ in $\mathbf{Z_{LR}}$ is generated by downsampling the high-resolution color image $\mathbf{X_{i,HR}}$ using bicubic interpolation followed by grayscale conversion.

In order to learn the mapping between a low-resolution grayscale image and a high-resolution color image, it is necessary to first upsample the low-resolution grayscale image:

$$\mathbf{Z_{i,LR}} \xrightarrow[\text{Grayscale}]{\text{Upsampling}} \mathbf{Z_{i,HR}} \quad (2)$$

where, $\mathbf{Z_{i,HR}}$ is the upsampled high-resolution grayscale image.

However, the upsampled image becomes blurred due to significant information loss. Inspired by the work of Isola et al. [7], the problem is overcome using a Pix2Pix generative adversarial network to learn the mapping between the upsampled grayscale image and the high-resolution color image.

Specifically, a conditional generative adversarial network may be used to learn the mapping from input images to the corresponding output images (i.e., image-to-image translation). In this work, it is required to learn the mapping from the input upsampled grayscale image set $\mathbf{Z_{HR}}$ to the output high-resolution color image set $\mathbf{X_{HR}}$. Therefore, the objective function is written as:

$$\mathcal{L}_{cGAN}(G, D) = \mathbb{E}_{\mathbf{Z_{HR}},\mathbf{X_{HR}}}[logD(\mathbf{Z_{HR}}, \mathbf{X_{HR}})] + \\ \mathbb{E}_{\mathbf{Z_{HR}},z}[log(1 - D(G(\mathbf{Z_{HR}}, z)))] \quad (3)$$

As mentioned above, a low-resolution image becomes blurred after upsampling. Therefore, the $L1$ norm (Manhattan distance) is employed to encode sharpness in the output images. The new objective function is given by:

$$\mathcal{L}_{cGAN}(G, D) = \mathbb{E}_{\mathbf{Z_{HR}}, \mathbf{X_{HR}}}[logD(\mathbf{Z_{HR}}, \mathbf{X_{HR}})] +$$
$$\mathbb{E}_{\mathbf{Z_{HR}}, z}[log(1 - D(G(\mathbf{Z_{HR}}, z)))] + \lambda\mathcal{L}_{L1}(G) \quad (4)$$

- **Image Matching:** The verification task is to determine if an image encoded as a quick response code matches the input image associated with the document or matches the image captured in real time.

Let $\mathbf{P_i}$ be the input image to be matched against the image encoded in the quick response code. For this purpose, the low-resolution grayscale image in the quick response code is extracted and mapped to the high-resolution color image $\mathbf{X_{i,HR}}$. Next, the two images are input to a pre-trained facial model to obtain the output facial representation vectors for the two images. The Euclidean distance between the two output facial representation vectors is computed to obtain the image matching score S_i, which is given by:

$$S_i = ||R(\mathbf{P_i}) - R(\mathbf{X_{i,HR}})||_F \qquad (5)$$

where $R(.)$ is the function that extracts facial features from the input images and $||.||_F$ is the Frobenius norm.

4.   Experiments and Results

The proposed method was evaluated using the Multi-PIE 51 dataset [5]. The evaluation employed 16×16 and 8×8 low-resolution images.

The Multi-PIE 51 dataset contains 50,248 images of 337 subjects. The images of each subject have differing illuminations, poses and expressions. The dataset was partitioned into a training set and testing set with 202 (60%) subjects and 135 (40%) subjects, respectively.

Two experiments were conducted: (i) image-to-image verification; and (ii) real-time person verification. Table 2 shows the details of the the experiments, which were conducted with 16×16 and 8×8 resolution images. The Light CNN-29 [17] and VGGFace [9] facial representation models were employed to extract features from images.

In order to encode images as quick response codes, the low-resolution images were converted to unicode and then stored as quick response codes. The steps were reversed to decode the images. The Pix2Pix generative adversarial network model was trained for 150 epochs using the 16×16 and 8×8 resolution images.

Table 2. Experiments conducted with 16×16 and 8×8 resolution images.

Experiment	Resolution	Training Samples	Testing Samples
Image-to-Image	16×16	33,613	16,635
Verification	8×8	33,613	16,635
Real-Time Person	16×16	33,613	16,635
Verification	8×8	33,613	16,635

4.1 Performance Evaluation

The proposed method was evaluated for image-to-image verification and real-time person verification. For both the evaluations, 16,635 genuine and imposter pairs were generated using test dataset. In the case of image-to-image verification, two same images (regardless of their resolutions) were considered to be a genuine pair. In the case of real-time person verification, two images of the same subject were considered to be a genuine pair.

4.2 Image-to-Image Verification

Image-to-image verification compared images in the documents against the images encoded as quick response codes in the documents. Figures 3 and 4 show the receiver operating characteristic (ROC) curves obtained for the Light CNN-29 and VGGFace models, respectively. Note that the curves on the left-hand sides of the figures are for 16×16 resolution images whereas the curves on the right-hand sides of the figures are for 8×8 resolution images. The ROC curves in the two figures indicate that image-to-image verification yields better results with the 16×16 resolution images for both the models. Also, the Light CNN-29 model performs better with 16×16 resolution images whereas the VGGFace performs better with 8×8 resolution images.

Table 3 shows the true positive rates for three false positive rates (0.01, 0.1 and 0.2) for the Light CNN-29 and VGGFace models with the two image resolutions. The first row of the table shows the image-to-image verification results. Note that the Light CNN-29 model yields a better true positive rate with 16×16 resolution images for a false positive rate of 0.01 whereas the VGGFace model yields better true positive rates for false positive rates of 0.1 and 0.2. In the case of 8×8 resolution images, the VGGFace model yields better true positive rates for all three false positive rates.

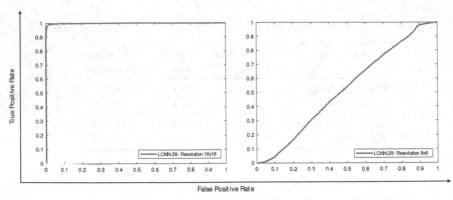

Figure 3. ROC curves for image-to-image verification (Light CNN-29 model).

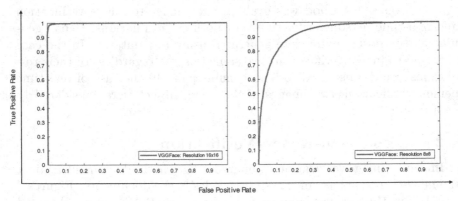

Figure 4. ROC curves for image-to-image verification (VGGFace model).

Table 3. True positive rates for false positive rates of 0.01, 0.1 and 0.2.

Experiment	Resolution	Light CNN-29			VGGFace		
		0.01	0.1	0.2	0.01	0.1	0.2
Image-to-Image	16 × 16	0.9823	0.9988	0.9996	0.9818	0.9992	0.9999
Verification	8 × 8	0.0461	0.0461	0.1642	0.3068	0.7857	0.9132
Real-Time Person	16 × 16	0.0598	0.6439	0.9396	0.0538	0.4582	0.6933
Verification	8 × 8	0.0087	0.0977	0.1476	0.0429	0.3070	0.4864

Figure 5 shows image samples obtained using the Pix2Pix generative adversarial network model (i.e., mapped from low-resolution grayscale images to high-resolution color images). The first row shows the original images, the second row shows the 16 × 16 resolution images encoded as quick response codes and the third row shows the images obtained

Figure 5. Images generated using Pix2Pix for 16×16 resolution images.

using Pix2Pix. Note that the visual appearances of subjects are almost completely preserved in the 16×16 resolution images. This demonstrates that the proposed method is able to reproduce images that are similar to the original images.

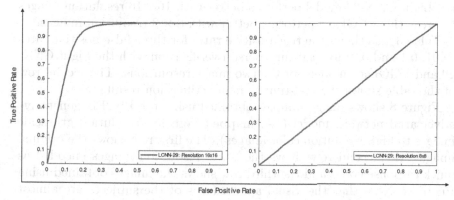

Figure 6. ROC curves for real-time person verification (Light CNN-29 model).

4.3   Real-Time Person Verification

Real-time person verification compared images encoded as quick response codes in the documents against persons' images captured in real time. Figures 6 and 7 show the receiver operating characteristic curves obtained for the Light CNN-29 and VGGFace models, respectively. Once again, the curves on the left-hand sides of the figures are for 16×16 resolution images whereas the curves on the right-hand sides of the figures

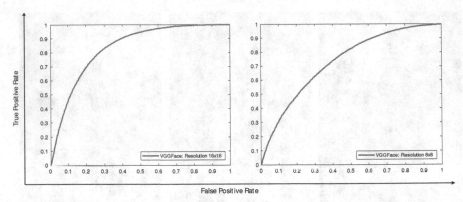

Figure 7. ROC curves for real-time person verification (VGGFace model).

are for 8×8 resolution images. The ROC curves reveal significant drops in real-time person verification compared with image-to-image verification – these are due to variations in illumination, poses and expressions in the images. The ROC curves for both models indicate that real-time person verification is better for 16×16 resolution images. Furthermore, the Light CNN-29 model performs better with 16×16 resolution images whereas the VGGFace performs better with 8×8 resolution images.

Table 3 also shows the true positive rates for three false positive rates (0.01, 0.1 and 0.2) for real-time person verification with the Light CNN-29 and VGGFace models for the two image resolutions. The second row of the table shows the real-time person verification results.

Figure 8 shows image samples obtained using the Pix2Pix generative adversarial network model (i.e., mapped from low-resolution grayscale images to high-resolution color images). The first row shows the original images, the second row shows the 8×8 resolution images encoded as quick response codes and the third row shows the images obtained using Pix2Pix. Note that the visual appearances of the subjects are almost completely distorted in all the images. This demonstrates that the 8×8 resolution images are not suitable for real-time person verification.

5.   Conclusions

The availability of sophisticated, yet inexpensive, printing and scanning equipment makes it easy for fraudsters to tamper with facial images on important documents such as identity cards, drivers' licenses, admit cards, examination hall tickets and academic transcripts. Several solutions have been developed to verify the authenticity of identity documents, but they are usually limited to verifying document original-

Figure 8. Images generated using Pix2Pix for 8 × 8 resolution images.

ity. Moreover, their underlying techniques often fail when applied to printed versions of identity documents. Some researchers have developed techniques for detecting the tampering of text in photocopied and printed versions of identity documents, but they do not verify the facial identities of persons from the documents.

This chapter has presented an effective and low-cost solution for verifying facial images from original and printed versions of identity documents. A person's facial image is converted to a quick response code that is embedded in an identity document during its creation. In image-to-image verification, the image on an identity document is compared against the image encoded as a quick response code on the document. In real-time person verification, the image encoded as a quick response code on an identity document is compared against the person's facial image captured in real time.

Future work will implement the proposed method in real-world environments. Also, research will focus on detecting fraudulent identity documents where the facial features of two persons are amalgamated to create facial images.

References

[1] S. Aygun and M. Akcay, Securing biometric face images via steganography for QR code, *Proceedings of the Eighth International Conference on Information Security and Cryptology*, pp. 128–133, 2015.

[2] S. Chhabra, G. Gupta, M. Gupta and G. Gupta, Detecting fraudulent bank checks, in *Advances in Digital Forensics XIII*, G. Peterson and S. Shenoi (Eds.), Springer, Cham, Switzerland, pp. 245–266, 2017.

[3] A. Espejel-Trujillo, I. Castillo-Camacho, M. Nakano-Miyatake and H. Perez-Meana, Identity document authentication based on VSS and QR codes, *Procedia Technology*, vol. 3, pp. 241–250, 2012.

[4] I. Goodfellow, J. Pouget-Abadie, M. Mirza, B. Xu, D. Warde-Farley, S. Ozair, A. Courville and Y. Bengio, Generative adversarial nets, *Proceedings of the Twenty-Seventh Annual Conference on Neural Information Processing Systems*, pp. 2672–2680, 2014.

[5] R. Gross, I. Matthews, J. Cohn, T. Kanade and S. Baker, Multi-PIE, *Image and Vision Computing*, vol. 28(5), pp. 807–813, 2010.

[6] G. Gupta, S. Saha, S. Chakraborty and C. Mazumdar, Document frauds: Identification and linking fake documents to scanners and printers, *Proceedings of the International Conference on Computing: Theory and Applications*, pp. 497–501, 2007.

[7] P. Isola, J. Zhu, T. Zhou and A. Efros, Image-to-Image Translation with Conditional Adversarial Networks, arXiv:1611.07004 (`arxiv.org/abs/1611.07004`), 2018.

[8] J. Nayak, S. Singh, S. Chhabra, G. Gupta, M. Gupta and G. Gupta, Detecting data leakage from hard copy documents, in *Advances in Digital Forensics XIV*, G. Peterson and S. Shenoi (Eds.), Springer, Cham, Switzerland, pp. 111–124, 2018.

[9] O. Parkhi, A. Vedaldi and A. Zisserman, Deep face recognition, *Proceedings of the British Machine Vision Conference*, pp. 41.1–41.12, 2015.

[10] A. Rai, I have been asked to shut my mouth, but work will go on – An interview with the whistleblower who exposed Madhya Pradesh Vyapam scam, *The News Minute*, February 25, 2015.

[11] N. Raval, A. Machanavajjhala and L. Cox, Protecting visual secrets using adversarial nets, *Proceedings of the IEEE Conference on Computer Vision and Pattern Recognition Workshops*, pp. 1329–1332, 2017.

[12] S. Sarkar, R. Verma and G. Gupta, Detecting counterfeit currency and identifying its source, in *Advances in Digital Forensics IX*, G. Peterson and S. Shenoi (Eds.), Springer, Berlin Heidelberg, Germany, pp. 367–384, 2013.

[13] V. Seenivasagam and R. Velumani, A QR code based zero-watermarking scheme for authentication of medical images in teleradiology cloud, *Computational and Mathematical Methods in Medicine*, article no. 516465, 2013.

[14] I. Tkachenko, W. Puech, C. Destruel, O. Strauss, J. Gaudin and C. Guichard, Two-level QR code for private message sharing and document authentication, *IEEE Transactions on Information Forensics and Security*, vol. 11(3), pp. 571–583, 2016.

[15] P. Tyre, How sophisticated test scams from China are making their way into the U.S., *The Atlantic*, March 21, 2016.

[16] M. Warasart and P. Kuacharoen, Paper-based document authentication using digital signature and QR code, *Proceedings of the Fourth International Conference on Computer Engineering and Technology*, pp. 94–98, 2012.

[17] X. Wu, R. He, Z. Sun and T. Tan, A light CNN for deep face representation with noisy labels, *IEEE Transactions on Information Forensics and Security*, vol. 13(11), pp. 2884–2896, 2018.

Chapter 11

USING NEURAL NETWORKS FOR FAKE COLORIZED IMAGE DETECTION

Yuze Li, Yaping Zhang, Liangfu Lu, Yongheng Jia and Jingcheng Liu

Abstract   Modern colorization techniques can create artificially-colorized images that are indistinguishable from natural color images. As a result, the detection of fake colorized images is attracting the interest of the digital forensics research community. This chapter tackles the challenge by introducing a detection approach that leverages neural networks. It analyzes the statistical differences between fake colorized images and their corresponding natural images, and shows that significant differences exist. A simple, but effective, feature extraction technique is proposed that utilizes cosine similarity to measure the overall similarity of normalized histogram distributions of various channels for natural and fake images. A special neural network with a simple structure but good performance is trained to detect fake colorized images. Experiments with datasets containing fake colorized images generated by three state-of-the-art colorization techniques demonstrate the performance and robustness of the proposed approach.

Keywords: Image forensics, fake colorized image detection, neural networks

1. Introduction

Digital image forensics is the process of collecting, identifying, analyzing and presenting evidence derived from digital image resources [2, 6]. Rapid advancements in image tampering techniques have made it increasingly difficult to distinguish between natural and fake images. Farid [6] divides image tampering techniques into six categories: (i) compositing; (ii) morphing; (iii) re-touching; (iv) enhancing; (v) computer-generating; and (vi) painting. While these categories cover most image tampering techniques, other more specific image tampering techniques such as colorization [13] and splicing [1, 5] have been proposed.

© IFIP International Federation for Information Processing 2019
Published by Springer Nature Switzerland AG 2019
G. Peterson and S. Shenoi (Eds.): Advances in Digital Forensics XV, IFIP AICT 569, pp. 201–215, 2019.
https://doi.org/10.1007/978-3-030-28752-8_11

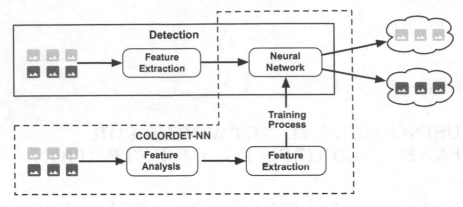

Figure 1. ColorDet-NN approach.

Colorization is the process of transforming grayscale images to colorized images by adding color features. Colorization techniques are frequently used to add color to greyscale photographs or black-and-white films to restore historical scenes. These techniques are also used to colorize black-and-white CT, X-ray and MRI images to enhance medical diagnosis and treatment. However, colorization can also be used for malicious purposes, for example, to create doctored photographs and videos that appear legitimate to the naked eye.

Guo et al. [8] were among the first researchers to focus on fake colorized image detection. They proposed two classification techniques, FCID-HIST and FCID-FE, that rely on support vector machines [3]. Difficulties in choosing appropriate kernel functions for the support vector machines limit the performance of the techniques. Additionally, the computing resources required by support vector machines render them infeasible for large datasets.

To address these challenges, this research employs the ColorDet-NN neural network [15] to detect fake colorized images. Figure 1 shows the ColorDet-NN approach. An initial feature analysis step compares the statistical differences in the color distributions of natural images and fake colorized images. A feature extraction step then captures contributing features from raw image data for pre-processing. The final training step utilizes the extracted features to create a ColorDet-NN neural network that detects fake colorized images.

2.   Background

Colorization is the process of transforming grayscale images to colorized images by adding color features. Several colorization techniques have been proposed over the past two decades. The colorization techniques differ in how they obtain and handle the data used to model the correspondence between grayscale and colorized images. As a result, colorization techniques are broadly divided into three categories: (i) scribble-based; (ii) transfer-based; and (iii) fully automated.

Scribble-based methods require users to specify the colors in grayscale images in advance based on their experience. The first scribble-based method, developed by Levin et al. [16], utilizes a quadratic cost function of the differences between a pixel and its neighboring pixels under the assumption that adjacent pixels with similar intensities should have similar colors. Several researchers have developed more effective techniques. For example, Luan et al. [17] have developed an interactive system for colorizing natural images that uses texture similarity to obtain effective color propagation. Sykora et al. [20] have created a flexible, interactive tool for painting hand-drawn cartoons. However, these techniques rely on – and are therefore limited by – the user's experience, and require a large number of experiments to achieve good performance.

Transfer-based colorizing techniques establish mappings between reference colorized images and grayscale images, following which they transfer colors to the target greyscale images from analogous regions of the reference colorized images. Reinhard et al. [18] have done pioneering research on transferring colors between images. Ironi et al. [12] have presented a novel color transfer technique that analyzes the low-level feature space using a robust supervised classification scheme. However, in transfer-based colorization, the choice of appropriate reference colorized images is crucial to obtaining good performance.

Several researchers have applied deep learning techniques [14] to colorization. These fully-automated techniques have better performance than scribble-based and transfer-based methods. Larsson et al. [13] have developed a fully-automated image colorization technique that predicts per-pixel color histograms utilizing low-level and semantic representations. Iizuka et al. [11] have employed a neural network that combines global priors and local image features to automatically colorize grayscale images. Zhang et al. [22] have proposed a fully-automated technique that increases the diversity of colors in images by posing colorization as a classification problem.

Guo et al. [8] were among the first researchers to leverage machine learning to detect fake colorized images. They proposed two classification methods, FCID-HIST and FCID-FE, that compute the statistical differences in the hue, saturation, dark and bright channels in different ways; they then employ support vector machines to distinguish between natural and fake colorized images.

Table 1. Maximum absolute differences for natural and fake image distributions.

Colorization Technique	Red	Green	Blue	Hue	Saturation	Value
Larsson et al. [13]	70%	42%	188%	264%	4,393%	174%
Iizuka et al. [11]	32%	211%	92%	1,093%	1,654%	52%
Zhang et al. [22]	105%	154%	70%	774%	5,083%	49%

3.　　Detection Methodology

Research in deep learning has significantly enhanced colorization techniques. It has become very difficult for humans to distinguish fake colorized images from natural images. The proposed ColorDet-NN approach for detecting fake colorized images effectively analyzes the statistical differences between natural images and fake colorized images generated by three state-of-the-art techniques developed by: (i) Larsson et al. [13]; (ii) Iizuka et al. [11]; and (iii) Zhang et al. [22]. The ColorDet-NN neural network is then trained to detect fake colorized images.

3.1　　Statistical Analysis and Testing

Statistical differences exist in the color distributions of natural images and fake colorized images. The RGB color space is defined by three chromaticities of the red, green and blue primary color channels (range is from 0 to 255), which can produce any chromaticity in the triangle defined by the primary colors. The HSV color space is an alternative representation of the RGB color space, which has hue, saturation and value channels. The RGB color space has more redundant information, which leads to insufficient feature differentiation. Therefore, the HSV color space is employed to obtain more features.

Normalized histograms were computed for the red, green, blue, hue, saturation and value channels in 10,000 natural images from the ImageNet LSVRC 2012 Validation Set [19]. The corresponding fake colorized images were generated using the colorization techniques of Larsson et al. [13], Iizuka et al. [11] and Zhang et al. [22].

The absolute differences between the distribution values of natural images and those of fake colorized images divided by the distribution values of the natural images were computed for the red, green, blue, hue, saturation and value channels. Table 1 shows the maximum values of the percentages obtained for the six channels. Clearly, a statistical difference exists in each channel between the natural and fake colorized images generated by each of the three colorization techniques.

Table 2. Two-sample Kolmogorov-Smirnov test results for the RGB channels.

Colorization Technique	Red	Green	Blue
Larsson et al. [13]	1	1	0
Iizuka et al. [11]	1	1	1
Zhang et al. [22]	0	0	1

Table 3. Two-sample Kolmogorov-Smirnov test results for the HSV channels.

Colorization Technique	Hue	Saturation	Value
Larsson et al. [13]	1	1	1
Iizuka et al. [11]	1	1	1
Zhang et al. [22]	0	1	0

Note that significant differences exist in the saturation channel. In this channel, all the percentages are more than 1,600%, which means that significant color biases exist at some channel values between the natural and fake colorized images. In addition, the minimum percentage reached 32%, which suggests that there are statistical differences that can be utilized for detection.

The two-sample Kolmogorov-Smirnov test [7] is employed to determine whether the distributions of natural images and fake colorized images are different. The test checks whether the two data samples have the same distributions in order to measure their differences.

The null hypothesis H_0 is defined as:

H_0 : The two data samples satisfy the same distribution.

Let $KSTest_m^c$ be the two-sample Kolmogorov-Smirnov test result between the distribution of natural images and the distribution of fake colorized images generated by a colorization method m in channel c. Then, the null hypothesis is rejected at the 0.05 level of significance if $KSTest_m^c = 1$.

Tables 2 and 3 show that at least one channel will reject the null hypothesis for each colorization method in each color space. In the case of fake colorized images generated using the technique of Iizuka et al. [11], the red, green, blue, hue, saturation and value channels all reject the null hypothesis. On the other hand, for fake colorized images generated using the technique of Zhang et al. [22], only the blue and saturation channels reject the null hypothesis, but this still means that the features of at least two channels can be used to distinguish between natural and fake colorized images. Simply put, there are statistical differences in the color distributions of natural and fake colorized images.

3.2   Feature Extraction

The statistical differences in the red, green, blue, hue, saturation and value channels are used for feature extraction.

Specifically, to distinguish between natural and fake colorized images the following six features are employed: (i) red channel feature F_r; (ii) green channel feature F_g; (iii) blue channel feature F_b; (iv) hue channel feature F_h; (v) saturation channel feature F_s; and (vi) value channel feature F_v.

Each channel feature is computed in a similar manner. For each feature F_{ch}, let $Hist_{n,ch}^{total}$ denote the normalized histogram distribution for all the natural images for channel ch. Let $Hist_{ch}^{\alpha}$ denote the ch channel histogram distribution for an input image α. The feature computation also leverages the first-order derivative of the normalized channel histogram distributions. These first-order derivatives are $Deri_{n,ch}^{total}$ for natural images. The $Deri_{ch}h^{\alpha}$ for an input image α is computed as:

$$Deri_{ch}^{\alpha}(i) = Hist_{ch}^{\alpha}(i+1) - Hist_{ch}^{\alpha}(i), \ i \in [0, 254] \qquad (1)$$

where $Hist_{ch}^{\alpha}(i)$ and $Deri_{ch}^{\alpha}(i)$ are components of the vectors $Hist_{ch}^{\alpha}$ and $Deri_{ch}^{\alpha}$, respectively.

Since a natural image has a closer similarity to the natural image distributions than a fake colorized image, the cosine similarity cos is used to measure the overall similarity between $Hist_{ch}^{\alpha}$ and $Hist_{n,ch}^{total}$ and $Deri_{ch}^{\alpha}$ and $Deri_{n,ch}^{total}$:

$$cos(\theta) = \frac{A \cdot B}{\|A\| \|B\|} = \frac{\sum_{i=0}^{n} A_i B_i}{\sqrt{\sum_{i=0}^{n} A_i^2}\sqrt{\sum_{i=0}^{n} B_i^2}} \qquad (2)$$

where A_i and B_i are components of vectors A and B, respectively.

The feature computations F_r^{α}, F_g^{α}, F_b^{α}, F_h^{α}, F_s^{α} and F_v^{α} for input image α are given by:

$$F_{ch}^{\alpha}(1) = \frac{Hist_{n,ch}^{total} \cdot Hist_{ch}^{\alpha}}{\left\| Hist_{n,ch}^{total} \right\| \left\| Hist_{ch}^{\alpha} \right\|}$$

$$= \frac{\sum_{i=0}^{255} Hist_{n,ch}^{total}(i) \times Hist_{ch}^{\alpha}(i)}{\sqrt{\sum_{i=0}^{255} Hist_{n,ch}^{total}(i)^2} \times \sqrt{\sum_{i=0}^{255} Hist_{ch}^{\alpha}(i)^2}}, \tag{3}$$

$$ch = r, g, b, h, s, v$$

$$F_{ch}^{\alpha}(2) = \frac{Deri_{n,ch}^{total} \cdot Deri_{ch}^{\alpha}}{\left\| Deri_{n,ch}^{total} \right\| \left\| Deri_{ch}^{\alpha} \right\|}$$

$$= \frac{\sum_{i=0}^{254} Deri_{n,ch}^{total}(i) \times Deri_{ch}^{\alpha}(i)}{\sqrt{\sum_{i=0}^{254} Deri_{n,ch}^{total}(i)^2} \times \sqrt{\sum_{i=0}^{254} Deri_{ch}^{\alpha}(i)^2}}, \tag{4}$$

$$ch = r, g, b, h, s, v$$

$$F_{ch}^{\alpha} = [F_{ch}^{\alpha}(1), F_{ch}^{\alpha}(2)], \ ch = r, g, b, h, s, v \tag{5}$$

After all the features are obtained, the feature vector F_{HIST}^{α} for an input image α is:

$$F_{HIST}^{\alpha} = [F_r^{\alpha}, F_g^{\alpha}, F_b^{\alpha}, F_h^{\alpha}, F_s^{\alpha}, F_v^{\alpha}] \tag{6}$$

Let L_{HIST}^{α} denote the binary label of F_{HIST}^{α}. L_{HIST}^{α} has a value of one if input image α is a fake colorized image and L_{HIST}^{α} has a value of zero if input image α is a natural image.

Thus, the final detection data D_{HIST}^{α} is:

$$D_{HIST}^{\alpha} = [F_{HIST}^{\alpha}, L_{HIST}^{\alpha}]$$

$$= \begin{cases} [F_r^{\alpha}, F_g^{\alpha}, F_b^{\alpha}, F_h^{\alpha}, F_s^{\alpha}, F_v^{\alpha}, 1], & \text{if image } \alpha \text{ is fake} \\ [F_r^{\alpha}, F_g^{\alpha}, F_b^{\alpha}, F_h^{\alpha}, F_s^{\alpha}, F_v^{\alpha}, 0], & \text{if image } \alpha \text{ is natural} \end{cases} \tag{7}$$

3.3   Neural Network Construction

An artificial neural network is an algorithm that models computations using graphs of artificial neurons, mimicking how neurons work in the brain. Artificial neural networks are well-suited to solving complex non-linear problems. Unlike traditional machine learning algorithms such as support vector machines, artificial neural networks have flexible structures that can be adapted according to the problem that is to be solved. This work uses an artificial neural network to differentiate natural images from fake colorized images.

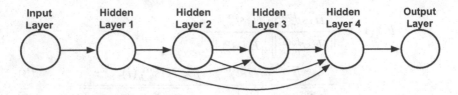

Figure 2.   Neural network structure.

The artificial neural network employed for detecting fake colorized images is based on the dense convolutional network (DenseNet) model [10]. DenseNet has a relatively simple structure, in which every layer of the network is connected to every other layer in a feed-forward manner. Compared with other neural network models, DenseNet strengthens feature propagation while reducing the number of parameters.

Figure 2 shows the structure of the neural network used for fake colorized image detection. The neural network has six layers – an input layer, an output layer and four hidden layers. Each hidden layer is fully connected to the previous layers. For each hidden layer, the input of the layer is the sum of the outputs of the other hidden layers.

The relationships of the hidden layers are given by:

$$X_i = Y_1 + \cdots + Y_{i-1}, \; i \geq 2 \tag{8}$$

where X_i and Y_i are the input and output of layer i, respectively.

The selection of an appropriate activation function is an important aspect when designing a neural network. The proposed technique employs a parametric rectified linear unit (PReLU) [9], an activation function with parameters that can be trained. This activation function is used in the hidden layers of the network. Table 4 shows the details of the neural network. Hidden layers 1 through 3 have 32 neurons each whereas hidden layer 4 has 128 neurons.

The joint supervision of the softmax loss function and center loss function [21] was used to train the neural network. The softmax loss function is one of the most widely used loss functions. The center loss function has been demonstrated to minimize intra-class variations while keeping the features of different classes separable.

The softmax loss function L_S is:

$$L_S = - \sum_{i=1}^{m} \log \frac{e^{W_{y_i}^T x_i + b_{y_i}}}{\sum_{j=1}^{n} e^{W_j^T x_i + b_j}} \tag{9}$$

where x_i is the ith deep feature, which belongs to the class y_i; m is the mini-batch; and n is the number of classes.

Table 4. Neural network details.

Layer	Neurons	Activation	Connected Layer
Input Layer	12		
Hidden Layer 1	32	PReLU	Input Layer
Hidden Layer 2	32	PReLU	Hidden Layer 1
Hidden Layer 3	32	PReLU	Hidden Layer 1, 2
Hidden Layer 4	128	PReLU	Hidden Layer 1, 2, 3
Output Layer (Softmax loss)	2		Hidden Layer 4
Output Layer (Center loss)	1		Hidden Layer 4

The center loss function is:

$$L_C = \frac{1}{2} \sum_{i=1}^{m} \|x_i - c_{y_i}\|_2^2 \tag{10}$$

where c_{y_i} is the center of y_i of the deep feature and is updated as the deep feature changes. The joint supervision of the softmax loss function and center loss function are used to train the neural network.

The final loss function is:

$$L = L_S + \lambda L_C$$
$$= -\sum_{i=1}^{m} \log \frac{e^{W_{y_i}^T x_i + b_{y_i}}}{\sum_{j=1}^{n} e^{W_j^T x_i + b_j}} + \frac{\lambda}{2} \sum_{i=1}^{m} \|x_i - c_{y_i}\|_2^2 \tag{11}$$

4. Experiments and Results

This section describes the datasets used in the experiments, the experimental measurements and the performance evaluation results.

4.1 Datasets

Six benchmark datasets based on the ImageNet LSVRC 2012 Validation Set [19] were employed in the experiments. The datasets, which are widely used in image colorization and fake image detection research, contain many categories of images, including images of people, animals, buildings and landscapes.

The D1 dataset corresponds to the ctest10k dataset [13], which has 10,000 fake colorized images and their corresponding 10,000 natural images from the ImageNet LSVRC 2012 Validation Set. Datasets D2 and D3 each contain the 10,000 natural images in dataset D1 as well as 10,000 fake colorized images generated from the natural images using the colorization techniques of Iizuka et al. [11] and Zhang et al. [22], respectively. Thus, datasets D1, D2 and D3 each have 20,000 images.

The D4 dataset contains 2,000 fake colorized images randomly selected from the ctest10k dataset [13] and their corresponding 2,000 natural images from the ImageNet LSVRC 2012 Validation Set, resulting in a total of 4,000 images. The D5 dataset also has 4,000 images – 2,000 natural images selected randomly from dataset D1 and their corresponding fake colorized images generated by the colorization technique of Iizuka et al. [11]. The D6 dataset also has 4,000 images – 2,000 natural images selected randomly from dataset D1 and their corresponding fake images generated by the colorization technique of Zhang et al. [22].

4.2   Measurements

The accuracy, precision, recall and F1 score were used to evaluate the performance of ColorDet-NN. In addition, the half total error rate (HTER) and area under the curve (AUC) measurements were used to compare the performance of ColorDet-NN against the performance of FCID-HIST and FCID-FE developed by Guo et al. [8].

4.3   Performance Evaluation

Several experiments were designed to evaluate the performance of ColorDet-NN. The experiments use all six datasets, D1 through D6.

The first set of experiments evaluated the ability of ColorDet-NN to detect fake colorized images. Datasets D1, D2 and D3 were used to assess the performance of ColorDet-NN at detecting fake colorized images generated using the colorization techniques of Larsson et al. [13], Iizuka et al. [11] and Zhang et al. [22]. Each dataset D1, D2 and D3 was randomly divided into a training set corresponding to 75% of the dataset and a testing set corresponding to 25% of the dataset.

Nine cross-validation experiments were conducted using the three training sets and three testing sets. The results in Table 5 demonstrate that ColorDet-NN can effectively distinguish between natural images and the fake colorized images generated by the three colorization techniques. All the accuracy values are greater than 88% when the training and testing sets come from the same original dataset. However, the accuracy values fall when the training and testing sets come from different datasets. Most of the experiments have accuracy values greater than 73%, except for the third experiment; this is likely due to large differences in the image features for fake images generated by the colorization techniques.

Table 5. Detection results in the cross-validation experiments.

Training	Testing	Accuracy	Precision	Recall	F1 Score	HTER
D1	D1	88.46%	85.46%	92.67%	88.92%	11.54%
	D2	74.32%	74.24%	74.48%	74.36%	25.68%
	D3	65.94%	69.17%	57.52%	62.81%	34.06%
D2	D1	77.82%	84.52%	68.12%	75.44%	22.18%
	D2	88.20%	87.59%	88.99%	88.28%	11.80%
	D3	80.82%	87.46%	71.96%	78.96%	19.18%
D3	D1	73.76%	78.29%	65.76%	71.48%	26.24%
	D2	81.16%	80.22%	82.72%	81.45%	18.84%
	D3	89.58%	86.28%	94.12%	90.02%	10.42%

Table 6. Area under curve results in the cross-validation experiments.

Training	Testing		
	D1	D2	D3
D1	0.95972	0.82466	0.73605
D2	0.87773	0.95323	0.89848
D3	0.82149	0.89930	0.96504

Table 6 shows the area under the curve results in the cross-validation experiments. All the area under the curve results are greater than 95% when the training and testing sets come from the same dataset. The results imply that ColorDet-NN is effective at detecting the fake colorized images.

The next set of experiments were conducted to compare the detection performance of ColorDet-NN against state-of-the-art techniques for detecting fake colorized images. The FCID-HIST and FCID-FE fake colorized image detection techniques developed by Guo et al. [8] were used in the comparisons. Datasets D4, D5 and D6 were divided equally into training sets and testing sets in order to evaluate the performance of ColorDet-NN versus FCID-HIST and FCID-FE.

Nine experiments were performed using testing and training sets drawn from the same and different datasets. Table 7 compares the area under the curve results for ColorDet-NN, FCID-HIST and FCID-FE. ColorDet-NN has better performance than FCID-HIST and FCID-FE in most situations, especially when the training and testing sets are drawn from the same dataset (area under the curve values greater than 93%). A

Table 7.   Comparison of area under the curve results.

Training	Testing	ColorDet-NN	FCID-HIST	FCID-FE
D4	D4	0.93654	0.85687	0.85762
	D5	0.81590	0.74112	0.77320
	D6	0.76650	0.78525	0.84100
D5	D4	0.83701	0.63106	0.68758
	D5	0.93613	0.85927	0.90382
	D6	0.88755	0.56928	0.71286
D6	D4	0.82165	0.79804	0.82669
	D5	0.88527	0.63953	0.72447
	D6	0.94972	0.83215	0.84832

small decline in performance is seen when the training and testing sets come from different datasets, but ColorDet-NN still outperforms FCID-HIST and FCID-FE, except in the third and seventh experiments. The negative results in these two cases arise because dataset D6 has more complex features than dataset D4 and the features extracted by FCID-HIST and FCID-FE are more sensitive than the features extracted by ColorDet-NN.

Table 8.   Comparison of half total error rate results.

Training	Testing	ColorDet-NN	FCID-HIST	FCID-FE
D4	D4	13.85%	22.50%	22.30%
	D5	27.00%	33.95%	31.70%
	D6	30.45%	28.00%	23.65%
D5	D4	25.45%	38.15%	38.50%
	D5	13.85%	22.35%	17.30%
	D6	20.95%	43.55%	36.15%
D6	D4	25.80%	26.95%	25.10%
	D5	20.55%	41.85%	34.25%
	D6	12.35%	24.45%	22.85%

Table 8 compares the half total error rate results for ColorDet-NN, FCID-HIST and FCID-FE. ColorDet-NN has lower values than those of FCID-HIST and FCID-FE, which implies that ColorDet-NN outperforms FCID-HIST and FCID-FE in detecting fake colorized images.

In summary, the experiments demonstrate that ColorDet-NN has better performance than FCID-HIST and FCID-FE in distinguishing natural images from fake colorized images.

5.   Conclusions

The ColorDet-NN neural-network-based technique for detecting fake colorized images has three steps. The first step analyzes and validates the statistical differences existing between fake colorized images and their corresponding natural counterparts. The second step employs the cosine similarity of normalized histogram distributions between fake and natural images in various channels to extract features for detection. The third step designs and trains ColorDet-NN to detect fake colorized images. Experiments with six datasets containing fake colorized images generated by three state-of-the-art colorization techniques demonstrate that ColorDet-NN significantly outperforms existing detection methods.

The ColorDet-NN technique exhibits reduced performance when its training and testing sets are drawn from different datasets. This occurs because different colorization techniques with large differences in the statistical information of color distributions significantly impact the extraction of features used for fake image detection. Future research will focus on the common features of colorization techniques and leveraging auxiliary features such as texture to enhance detection. Additionally, efficient neural network structures will be investigated as a means to improve performance.

References

[1] K. Bahrami, A. Kot, L. Li and H. Li, Blurred image splicing localization by exposing blur type inconsistency, *IEEE Transactions on Information Forensics and Security*, vol. 10(5), pp. 999–1009, 2015.

[2] R. Bohme and M. Kirchner, Counter-forensics: Attacking image forensics, in *Digital Image Forensics*, H. Sencar and N. Memon (Eds.), Springer, New York, pp. 327–366, 2013.

[3] C. Chang and C. Lin, LIBSVM: A library for support vector machines, *ACM Transactions on Intelligent Systems and Technology*, vol. 2(3), article no. 27, 2011.

[4] H. Farid, Creating and Detecting Doctored and Virtual Images: Implications to The Child Pornography Prevention Act, Technical Report TR2004-518, Department of Computer Science, Dartmouth College, Hanover, New Hampshire, 2004.

[5] H. Farid, Exposing digital forgeries from JPEG ghosts, *IEEE Transactions on Information Forensics and Security*, vol. 4(1), pp. 154–160, 2009.

[6] H. Farid, Image forgery detection, *IEEE Signal Processing*, vol. 26(2), pp. 16–25, 2009.

[7] G. Fasano and A. Franceschini, A multidimensional version of the Kolmogorov-Smirnov test, *Monthly Notices of the Royal Astronomical Society*, vol. 225(1), pp. 155–170, 1987.

[8] Y. Guo, X. Cao, W. Zhang and R. Wang, Fake colorized image detection, *IEEE Transactions on Information Forensics and Security*, vol. 13(8), pp. 1932–1944, 2018.

[9] K. He, X. Zhang, S. Ren and J. Sun, Delving deep into rectifiers: Surpassing human-level performance on ImageNet classification, *Proceedings of the IEEE International Conference on Computer Vision*, pp. 1026–1034, 2015.

[10] G. Huang, Z. Liu, L. van der Maaten and K. Weinberger, Densely connected convolutional networks, *Proceedings of the IEEE Conference on Computer Vision and Pattern Recognition*, pp. 2261–2269, 2017.

[11] S. Iizuka, E. Simo-Serra and H. Ishikawa, Let there be color! Joint end-to-end learning of global and local image priors for automatic image colorization with simultaneous classification, *ACM Transactions on Graphics*, vol. 35(4), article no. 110, 2016.

[12] R. Ironi, D. Cohen-Or and D. Lischinski, Colorization by example, *Proceedings of the Sixteenth Eurographics Conference on Rendering Techniques*, pp. 201–210, 2005.

[13] G. Larsson, M. Maire and G. Shakhnarovich, Learning representations for automatic colorization, *Proceedings of the Fourteenth European Conference on Computer Vision*, Part IV, pp. 577–593, 2016.

[14] Y. LeCun, Y. Bengio and G. Hinton, Deep learning, *Nature*, vol. 521(7553), pp. 436–444, 2015.

[15] Y. LeCun, B. Boser, J. Denker, D. Henderson, R. Howard, W. Hubbard and L. Jackel, Backpropagation applied to handwritten ZIP code recognition, *Neural Computation*, vol. 1(4), pp. 541–551, 1989.

[16] A. Levin, D. Lischinski and Y. Weiss, Colorization using optimization, *ACM Transactions on Graphics*, vol. 23(3), pp. 689–694, 2004.

[17] Q. Luan, F. Wen, D. Cohen-Or, L. Liang, Y. Xu and H. Shum, Natural image colorization, *Proceedings of the Eighteenth Eurographics Conference on Rendering Techniques*, pp. 309–320, 2007.

[18] E. Reinhard, M. Adhikhmin, B. Gooch and P. Shirley, Color transfer between images, *IEEE Computer Graphics and Applications*, vol. 21(5), pp. 34–41, 2001.

[19] O. Russakovsky, J. Deng, H. Su, J. Krause, S. Satheesh, S. Ma, Z. Huang, A. Karpathy, A. Khosla, M. Bernstein, A. Berg and F. Li, ImageNet Large-Scale Visual Recognition Challenge, *International Journal of Computer Vision*, vol. 115(3), pp. 211–252, 2015.

[20] D. Sykora, J. Dingliana and S. Collins, LazyBrush: Flexible painting tool for hand-drawn cartoons, *Computer Graphics Forum*, vol. 28(2), pp. 599–608, 2009.

[21] Y. Wen, K. Zhang, Z. Li and Y. Qiao, A discriminative feature learning approach for deep face recognition, *Proceedings of the Fourteenth European Conference on Computer Vision*, Part VII, pp. 499–515, 2016.

[22] R. Zhang, P. Isola and A. Efros, Colorful image colorization, *Proceedings of the Fourteenth European Conference on Computer Vision*, Part III, pp. 649–666, 2016.

V

FORENSIC TECHNIQUES

Chapter 12

DIGITAL FORENSIC ATOMIC FORCE MICROSCOPY OF SEMICONDUCTOR MEMORY ARRAYS

Struan Gray and Stefan Axelsson

Abstract   Atomic force microscopy is an analytical technique that provides very high spatial resolution with independent measurements of surface topography and electrical properties. This chapter assesses the potential for atomic force microscopy to read data stored as local charges in the cells of memory chips, with an emphasis on simple sample preparation ("delidding") and imaging of the topsides of chip structures, thereby avoiding complex and destructive techniques such as backside etching and polishing. Atomic force microscopy measurements of a vintage EPROM chip demonstrate that imaging is possible even when sample cleanliness, stability and topographical roughness are decidedly suboptimal. As feature sizes slip below the resolution limits of optical microscopy, atomic force microscopy offers a promising route for functional characterization of semiconductor memory structures in RAM chips, microprocessors and cryptographic hardware.

Keywords: Atomic force microscopy, memory chip delidding, surface imaging

1.   Introduction

Atomic force microscopy has been used to investigate the structures of memory devices and to conduct detailed failure analyses of memory cell structures. However, limited information is available about the use of atomic force microscopy to read the memory content of packaged chips. The published information suggests that atomic force microscopy and related techniques should work – the open question is how well.

The initial results presented in this chapter reveal that, even without custom sample mounting or modification of the atomic force microscope itself, it is possible to obtain topographic data from a packaged chip

© IFIP International Federation for Information Processing 2019
Published by Springer Nature Switzerland AG 2019
G. Peterson and S. Shenoi (Eds.): Advances in Digital Forensics XV, IFIP AICT 569, pp. 219–237, 2019.
https://doi.org/10.1007/978-3-030-28752-8_12

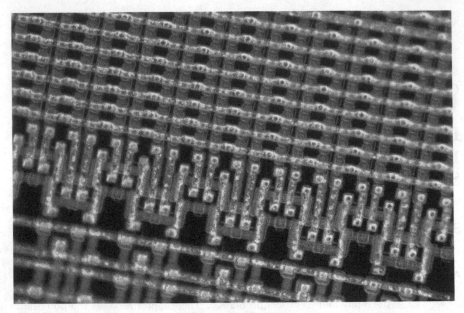

Figure 1. EPROM memory circuits imaged by dark-field optical microscopy.

using a basic research-grade instrument (Figure 1). Should future work prove the feasibility of the technique, it is easy to envision the creation of a custom atomic force microscope that could accommodate integrated circuits *in situ* on circuit boards. A key aspect of investigations in this area would be to perform topside imaging and characterization, avoiding the need for backside polishing and etching and, in principle, maintaining the integrity of the chip as a working device.

There are some interesting prospects for the future. Micro-electro-mechanical systems (MEMS) technology and other manufacturing processes, which could produce smaller, lighter atomic force microscopy structures with higher fundamental resonance, would enable an increase in data access rates and crash-free investigations of surfaces with high relief. These, in turn, would impact the practicability and security of the applications discussed in this chapter. Further development of high-speed electronics and microwave engineering may permit other advances in surface characterization of semiconductor devices or simply make measurements easier, cheaper and more reliable.

From the point of view of forensic investigations, atomic force microscopy offers a number of advantages: it is minimally invasive; it may be performed in a wide variety of environments; and it can be used to image almost any kind of sample. The problems in performing atomic force microscopy studies of working memory chips are mostly practical, but are of sufficient severity to deter casual investigations. Whatever the future may bring in terms of instruments and their capabilities, one thing is clear: sample preparation techniques will continue to be very important.

This chapter focuses on atomic force microscopy and related techniques, and how future developments could make them more applicable to forensic investigations of memory chips. The results of preliminary experiments are used to illuminate the practical issues that limit successful implementation.

2. Background

This section discusses probe microscopy and relevant issues related to security and forensics.

2.1 Probe Microscopy

Probe microscopy has antecedents in various stylus-based surface profiling tools, but the dramatic increases in sensitivity and resolution provided by the invention of scanning tunneling microscopy (STM) in 1981 [4] have led to the explosive development of instruments and their applications. Atomic force microscopy (AFM) was invented soon afterwards [3], motivated by a desire to expand the atomic resolution of the scanning tunneling microscope to investigations of non-conducting samples. Atomic force microscopy uses a force interaction between the probe tip and the sample surface to measure the distance between them. Typically, the interaction involves the Van der Waals force, but in principle, any force that varies with the relative positions of the tip and sample may be employed.

The narrow focus on atomic resolution as the ultimate goal of probe microscopy has ensured that the early literature in the field is full of interesting experiments and phenomena that failed to make an impact because of what was perceived as "poor resolution." Many of these techniques have been justifiably neglected, but some loiter at the margins of respectability and are worth revisiting periodically to examine whether or not they have acquired contemporary relevance. An example is using microwave or radio-frequency signals to measure the electrical characteristics of a sample surface. This was proved to be possible in the early days, but the methods were not robust enough to be adopted widely. However, novel measurements made recently using scanning microwave impedance microscopy (SMIM) have rejuvenated the field, revealing spatial variations in surface capacitance with nanometer resolution [15].

In addition to scanning microwave impedance microscopy, two more established techniques – electrical force microscopy (EFM) and scanning capacitance microscopy (SCM) – are relevant to forensic investigations of chip surfaces. All three techniques use the voltage on a conductive tip to reveal additional information about a surface beyond its topography. Electrical force microscopy measures the Coulomb forces between the tip and any charge concentrations on the sample surface. Scanning capacitance microscopy uses a modulated voltage to reveal changes in the tip-surface capacitance that are related to local doping levels, stored charge and metallization of semiconductor devices. Scanning microwave impedance microscopy investigates similar factors by regarding the tip-sample junction as the termination of a transmission line using the back-reflected signal to measure the complex tip-sample impedance.

2.2   Security and Forensics

Secure communications and computing have been important for many years and their importance seems set to increase. Several techniques have been devised to defend against attacks on confidentiality, integrity and availability. These include the use of cryptography [22], information flow analysis [23], and detection and estimation theory [1]. The vast majority of these techniques depend on a secure "black box" to hide the secrets or to perform computations. The concept takes different names depending on how it is implemented, including trusted computing bases in secure operating systems [20], bastion hosts in firewalls [19] and trusted platform modules in hardware-supported security [12].

In the case of secure operating systems and firewalls, many types of attacks are known, for example, those based on the exploitation of (inevitable) software flaws. Likewise, a number of attacks against hardware have been devised, many of them based on observing, or affecting, the hardware operating environment. The attacks include differential power analysis [13], differential fault analysis [2], probing with light/laser excitation [21], freezing of memory [10], electron microscope analysis [17] and Van Eck/TEMPEST radiation analysis. All these attacks, with one or two exceptions, require physical access to the hardware. Physical access is becoming increasingly easy to obtain as secure electronic hardware in the form of embedded and mobile devices is becoming commonplace. As these devices, especially those in mobile phones, have become ubiquitous, increased security requirements have led manufacturers to rely on hardware-based cryptographic modules and ubiquitous encryption to maintain the security of user data [9]. This has presented significant challenges to law enforcement and other actors who need to access the stored data in order to investigate crimes and other incidents. While there has been some success in leveraging software flaws in security implementations of smartphones, it is unclear how long this avenue will remain effective [8].

It follows that other techniques, including piercing the black box itself using sophisticated analysis techniques, are probably the inevitable next step in the evolution of digital forensics. However, it should be noted that knowledge pertaining to such attacks is not only useful to would-be attackers (e.g., law enforcement), but also to defenders, because it is difficult to defend against unknown threats. One such threat that should not be ignored is the question of whether or not the hardware can be trusted, especially if backdoors have been introduced during the manufacturing process [24].

3. Atomic Force Microscopy

Before examining more specialized techniques and how they have been applied to investigations of semiconductor memory devices, it is instructive to summarize the benefits and problems of atomic force microscopy because they are relevant to all derived technologies. Atomic force microscopy is a mature technique and the purpose here is not to reproduce the wealth of information in the numerous textbooks, manufacturer application notes and (surprisingly reliable) Wikipedia entries. Instead, the technique is briefly described with an emphasis on key performance metrics that are relevant to digital forensics as well as aspects that are currently limiting, but where future developments could have significant impact.

Figure 2 shows a generic atomic force microscopy setup in which a laser is reflected from the back of a cantilever carrying a sharp tip. A split photodiode measures the deflection of the laser beam, which changes as the cantilever flexes in response to the forces between the tip and the sample. The deflection signal is fed to a feedback loop that adjusts the position of the cantilever mount to keep the force and, hence, the height above the surface constant. As the tip moves sideways, its position tracks changes in the surface topography while the feedback loop maintains a constant height.

Most routine work has stabilized around the use of probes manufactured via semiconductor microfabrication techniques, with the tip and cantilever integrated in a standardized chip. The stiffness and other mechanical properties of the cantilever can be readily tuned; other aspects of the tip can also be optimized for particular applications. For example, conductive tips for the techniques discussed in this chapter can be

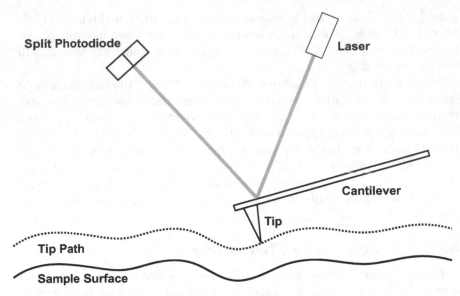

Figure 2. Atomic force microscopy.

made by doping the semiconductor material used in their manufacture or by evaporating metals onto the structure. Cantilevers typically have force constants ranging from 0.1 to 100 N/m depending on their intended modes of operation. Typical atomic force microscopy resolutions are 0.1 to 10 nm vertically (largely limited by noise) and 1 to 100 nm laterally (largely limited by tip shape, and tip and sample quality).

Sensing position via the static flexing of the cantilever – the so-called "contact" mode – is an option for most microscopes, but the forces between the tip and sample tend to be large, often leading to damage or wear. It is more common to situate the tip farther away from the surface, where the forces are weaker, using a modulation technique to recover sensitivity. Such oscillation-based schemes are robust and reliable; because the tip spends most of its time far from the surface, the potential for wear or damage are reduced substantially. A high-amplitude oscillation also ensures that strong interactions at close distances (e.g., adhesion or attraction to an absorbed water layer in ambient conditions) do not cause substantial dragging or friction, enabling rough and unpredictable surfaces to be measured more easily.

Oscillatory schemes also enable forces other than the intrinsic Van der Waals interaction to be distinguished, for example, by modulating them with a different frequency, or because they operate out of phase or have a different gradient with respect to the distance from the sample. Thus, when measuring Coulomb forces in electrical force microscopy, it is possible to modulate the voltage on the tip at a frequency well below that of the cantilever oscillation and, thus, use a second lock-in measurement to assess the electrostatic force separately.

The most widely-quoted figures of merit for atomic force microscopy and related techniques usually involve resolution. This, as mentioned above, can be something of a distraction. Even a desktop atomic force microscope operating in air at standard temperature and humidity can readily achieve 10 nm resolution, provided that the sample surface is clean and stable. Although gate lengths and some oxide thicknesses in semiconductor devices are approaching these dimensions, the lateral spacing of any likely two-dimensional storage structure is considerably larger and it is not likely that the spatial resolution limit of atomic force microscopy would present a significant obstacle to determining whether or not a given memory cell is charged.

Other aspects of atomic force microscopy are likely to be more significant than resolution, especially bandwidth and the time domain. There are three key timescales that govern how an atomic force microscope measurement may be made. Perhaps the least significant is the most fundamental: all electromagnetic and chemical interactions between the tip and surface operate on a timescale determined by the propagation of electromagnetic energy in the near field and the response times of electrons and other charges in the materials. Typically, this corresponds to frequencies in or beyond the visible spectrum (10^{14} Hz or higher), which implies a typical timescale of femtoseconds or less. Two firm conclusions can be made because the timescale is so much faster than any near-term developments in clock speed or phase-coherent detectors. First, any experiment that is devised is unlikely to be limited by the fundamental electromagnetic properties of the materials, even at terahertz frequencies. Second, if the response to higher frequency stimulation is of interest, then a modulation scheme or some sort of heterodyne detection would be needed to shift the signal into a measurable band.

The most significant timescale for practical experiments, including digital forensic uses, is the pixel clock: the rate at which an atomic force microscopy system can take individual pixels of the final image. This is surprisingly slow: typical times for a raster scan are 0.1 to 1 seconds for each line of data, leading to acquisition times of an hour or more even for low-resolution images. The slow response is set by the fundamental mechanical frequency of the microscope as a whole. This limits the response to positioning commands from the feedback loop and scan drivers because operating above resonance with an unknown mechanical phase shift quickly leads to unrecoverable damage to the sample and/or tip. The rigidity of the microscope is determined by factors related to its intended use and special instruments can achieve better performance by dispensing with easy tip exchange or the need to accommodate a wide range of sample sizes.

This factor is also the one that may be most amenable to change and where developments directly applicable to forensic imaging are expected. For example, efforts are underway to use micro-electro-mechanical systems technology to make the entire atomic force microscope exchangeable rather than just the tip [16]. Integrating the motion actuators, cantilever and tip in a monolithic package can make the mechanical loop between substrate and tip considerably smaller and stiffer, raising its fundamental frequency and enabling faster scanning while still under closed loop control.

Faster scan speeds will allow more rapid processes to be recorded and studied, as well as make scanning more reliable and tip crashes less likely. As discussed in the pilot study below, the slow response of current microscopes is not only frustrating, but it leads to poor quality data. Improving predictability and reducing the likelihood of damaging the sample under study will only benefit digital forensic investigations in search of presentable evidence.

The third important timescale is the fundamental oscillation frequency of the cantilever. This is adjustable based on design and materials, but the usual value lies in the 100 to 300 kHz range. At present, this is not regarded as a limitation. Other signals can be modulated at 1 to 50 kHz and still be safely below the oscillation frequency while remaining well above the pixel clock rate. There may, however, be difficulties in the future. If, as is highly desirable, scan speeds increase and raise the pixel clock rate, limited bandwidth will be available for intermediate frequency measurements in electrical force microscopy, scanning capacitance microscopy and scanning microwave impedance microscopy.

One final property of atomic force microscopy is of relevance. Unlike an electron microscope with its relatively high beam energy or even a visible light microscope that uses photon energies capable of breaking chemical bonds, an atomic force microscope has a very soft touch. It is possible to use forces so low that any sample sensitive to them would be impractical as a device. It is also possible to measure electrical characteristics with extremely low applied voltages, currents and fields. This makes atomic force microscopy excellent for non-destructive testing and also makes it difficult to implement countermeasures in the chips being investigated.

Figure 3. EPROM memory chip with exposed die (approx. 4 mm×8 mm in size).

4. Memory Chip Layout and Structure

Like most semiconductor integrated circuits, memory chips are created by cycles of lithography, deposition and reaction to produce fine-scale patterns of various materials on the surface of a single-crystal semiconductor wafer, usually silicon. The basic chip or die is brittle and sensitive to chemical and mechanical damage, so it is usually packaged before being used in a device. First, thin wires are bonded to the chip die to create more robust connections to metal pins than can be accomplished using conventional soldering techniques. Next, the die and cage are encapsulated. Polymer resins are now used for packaging; sintered ceramics used to be common and are still employed in some devices.

Figure 3 shows a 1,024 B Intel 2708 EPROM memory chip made in 1974. The exposed chip die is approximately 4 mm×8 mm in size. Despite its venerable age and low capacity, the chip illustrates the general characteristics that are still shared by modern high density devices. The quartz window that protects the chip die has been removed to show the structure beneath it. Because the circuits etched into the chip surface are so large, they are easy to see with a camera lens or microscope.

Figure 4 shows a close-up view of the die itself. The two large blocks are the actual memory locations, which are surrounded by control circuitry and address lines used to write and read data. The figure also shows the bonding pads to which thin bonding wires are attached.

Figure 4. Close-up view of the chip die showing two arrays of memory cells.

The first lesson, which is also relevant to modern, high-density surface-mount integrated circuits, is that the overall package is much larger than the die itself. The second is that a substantial 3D structure surrounds the planar circuits on the die; this means that access to the memory array has to take place through a tunnel cut in the packaging and between the arcs of the bonding wires. This is by no means trivial for probe microscopy because non-standard or extended mounting of the cantilever impacts imaging performance.

Figure 5 shows an optical microscope image of the individual memory arrays. Charge is stored in the oval structures and electrical connections are made to a single memory cell by addressing the appropriate combination of white and yellow/gold lead-in traces. The spacing between the bright white address lines is approximately $20\,\mu m$. Modern chips are more complex and have structures that are an order of magnitude smaller. Also, they often include more transistors and other active devices as part of the individual memory cells. However, the general layout of memory chips is very similar as is the packaging.

5.   Prior Art

The three atomic force microscopy techniques – electrical force microscopy, scanning capacitance microscopy and scanning microwave im-

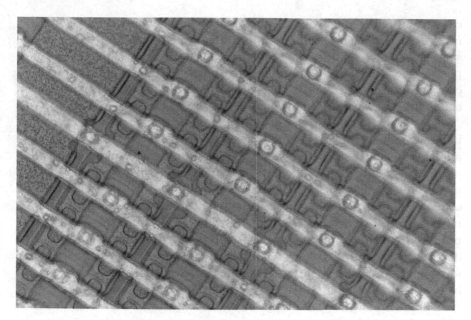

Figure 5. Optical microscope image of individual memory cells.

pedance microscopy – all have potential for investigations of semiconductor memory. They use a conductive tip to apply or sense electrical signals and typically employ conventional atomic force microscopy height sensing to control the tip position while measuring the electrical signals independently.

The first technique, electrical force microscopy, measures the electrical forces between a surface and a charged tip. An oscillating voltage is often applied to the tip and a lock-in amplifier is used to isolate this component of the signal coming from the split photodetector. Provided a frequency is chosen that lies between the pixel clock and the fundamental frequency of the cantilever, this perturbation neither affects the topographic image data nor the basic operation of the height stabilizing feedback. The lock-in signal represents the gradient of the electric force on the tip with distance. This signal changes sign when the tip is over positively or negatively charged regions of the surface. Although absolute quantitative measurements require detailed 3D models of the tip, surface geometry and materials, it is possible to map electric field gradients due to local charges.

The application to semiconductor memory devices is obvious because charges are mostly stored in capacitors or floating gate electrodes, which produce fringing fields that affect nearby sensing electrodes. Typical charge and discharge voltages are much larger than the tip voltages needed for good signal-to-noise in electrical force microscopy, so a measurement can be made without erasing the charge structure on the sample surface. Measuring the difference between a memory cell with a charged floating gate and a cell with a neutral gate is well within the capabilities of most instruments.

There is little reported work on using electrical force microscopy to image and read the contents of packaged semiconductor memory. An application note by Park Systems [18], a commercial manufacturer of atomic force microscopes, reports electrical force microscopy measurements on bare, unpackaged, uncontacted memory cells. Reports are also available about materials-science-oriented studies focused on developing novel memory structures (see, e.g., [6, 11]). However, these works do not describe successful top-down measurements of mainstream memory devices. Konopinski [14] has conducted the most extensive electrical force microscopy investigation of memory arrays, specifically flash memory EEPROMs used in SIM cards. While the electrical force microscopy measurements in this study were inconclusive, they do not rule out the utility of the technique.

The second technique, scanning capacitance microscopy, imposes a DC bias on the conducting tip with an overlaid oscillatory voltage. The currents on and off the tip are measured using a preamplifier placed close to the cantilever; a lock-in technique is used to provide specificity and noise rejection. The signal yielding the scanning capacitance microscopy measurement is proportional to dC/dV at the DC bias voltage. The technique is most applicable to systems where dC/dV varies with voltage, which includes many semiconductor structures.

Scanning capacitance microscopy has been used extensively to characterize on-chip transistor and memory structures in cross-sectional and top-down planar views. In fact, most manufacturers of commercial microscopes that offer scanning capacitance microscopy as an option provide an SRAM chip as a test sample. However, as in the case of the Park Systems technical note [18], the emphasis is usually on imaging a passivated planar sample with no actual connections to the doped regions of the semiconductor instead of *in situ* measurements of connected charged devices. The most attractive use of scanning capacitance microscopy is to assess dopant levels in semiconductors. It can reliably and robustly detect the difference between n-type and p-type regions. With suitable modeling, it is also able to measure doping levels as a function of position across the surface.

Some studies have used scanning capacitance microscopy to read the contents of memory cells. De Nardi et al. [7] have successfully used scanning capacitance microscopy to read memory devices that were extensively prepared for the scanning capacitance microscopy technique. They were able to distinguish between cells containing bit values of 1 and 0, and to recreate word-length data from scanning capacitance microscopy images. They also emphasize the need to map the physical locations of the data as seen by scanning capacitance microscopy to their logical locations within the conceptual data array. This is one area are where countermeasures such as scrambling physical memory locations on an individual chip could defeat read attempts. However, in cases where a consistent layout is used, the mapping can be performed for a single device (other than the device under test) and the results could be applied to all the devices in the same batch or of the same type.

The scanning capacitance microscopy study by De Nardi et al. [7] and the electrical force microscopy study by Konopinski [14] both involved the extraction of the die from the memory device and thinning it from the backside, reducing the chip thickness until the underside of the active circuits was almost exposed. This is not a trivial procedure, but is necessary, especially in the case of scanning capacitance microscopy, which relies on band-bending induced by the electric field from the tip to create a signal. In any case, it is essentially impractical to take measurements from the topside through the arrays of addressing and control lines.

The third technique, scanning microwave impedance microscopy [15], uses matched transmission lines and filters to apply a microwave signal to the conducting tip. A conventional RF network analyzer then records the back-reflected signal, essentially treating the tip-sample junction as an unmatched termination to the transmission line. From this, it is possible to extract the complex impedance of the junction and map it across the surface. As with scanning capacitance microscopy, the oscillatory signal enables the impedance measurement to be decoupled from the topography; the output signal is proportional to dC/dV for the imaginary part and to dR/dV for the real part. Scanning microwave impedance microscopy provides similar information as scanning capacitance microscopy, but with greater reliability and signal-to-noise. The published literature on scanning microwave impedance microscopy is a little thin, but the technique has definite promise.

6.   Vintage EPROM Chip Experiments

The experiments involved preparing and mounting a 1,024 B Intel 2708 EPROM memory chip in an atomic force microscope in order to investigate whether or not topside measurements are possible without the

Figure 6. Innova microscope with the EPROM chip mounted.

extensive and destructive sample preparation described in the electrical force microscopy and scanning capacitance microscopy studies discussed above.

The experiments employed an Innova atomic force microscope, a low-cost research-grade instrument, with scanning capacitance microscopy capabilities [5]. The atomic force microscopy system has a built-in video microscope that enables the tip position to be correlated with measurements using other microscopes. The system can perform atomic force microscopy on areas up to $100\,\mu\mathrm{m}\times100\,\mu\mathrm{m}$ in size.

Figure 6 shows the Innova atomic force microscope with its sound-proof cover removed and the EPROM chip mounted on the sample stage. The electronics for the scanning capacitance microscopy preamplifier are housed in the gold-colored rectangular box on the right.

Figure 7 shows a close-up view of the chip mounted *in situ* in the Innova atomic force microscope. The cantilever chip is mounted in the white holder at the top center. The red laser used for deflection detection can be seen reflecting off the cantilever itself, with some stray light to the left and on the surface of the EPROM chip. It is clear that space is extremely tight. Also, the rear of the atomic force microscope

Figure 7. Close-up view of the mounted EPROM chip.

scanner head is tilted up at an extreme angle so that the cantilever chip projects down into the well containing the chip die; this is not an optimal configuration.

Note that Figure 7 shows the EPROM chip in Figures 5 and 6 with its protective quartz window removed. The remainder of the packaging is still intact.

In order to investigate whether or not additional processing would facilitate atomic force microscopy, the top plate of the packaging was removed by cleaving it with a straight sharp blade. This enabled the microscope to be placed in a less contorted posture, but it was still not in its normal configuration. However, imaging was possible and atomic force microscope topographs could be taken.

Figure 8 shows an atomic force microscope topograph of the surface of the memory cell array. The image area is $20\,\mu\text{m} \times 20\,\mu\text{m}$ and the total height variation is $4.4\,\mu\text{m}$. The horizontal stripes and deep holes correspond to the bright white address lines seen in the optical image in Figure 5.

The topograph is remarkably clean and stable, although the uncleaned surface of a 1974-vintage EPROM chip was imaged. Moreover, the sur-

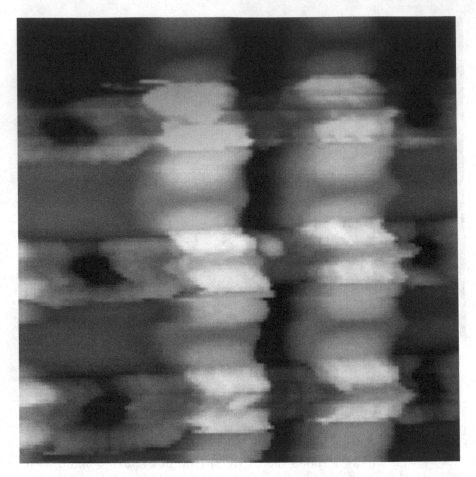

Figure 8. Topograph of the memory cell array in the EPROM chip.

face features can directly be related to those seen in the optical micro-
graph of the same chip shown in Figure 6. There is a great deal of tip
interaction, as can be seen from the sudden jumps and glitches at some
points in the image, but given the simplicity of the preparation technique, this is a very encouraging result. Note that although the package
had been delidded, all the bond wires were still intact and, in principle,
data could be written to the chip.

Experiments involving scanning capacitance microscopy and electrical
force microscopy were not conducted. However, it is clear that these
techniques would be entirely practicable, although delidding may be
more complicated for modern chips with resin packages. In fact, modern
chips with low-profile packaging and less surface relief on the die itself

should be easier to investigate because there would much less tip-sample interaction.

7.   Conclusions

Atomic force microscopy has already been used to investigate the structures of memory devices and to conduct detailed failure analyses of memory cell structures. However, limited information is available about its application to reading the memory contents of packaged chips. Nevertheless, the published research suggests that atomic force microscopy and related techniques should certainly be applicable to reading memory – the only question is how well.

The initial results presented in this chapter demonstrate that, even without custom sample mounting or modification of the atomic force microscope itself, it is possible to obtain topographic data from a packaged chip using a basic research-grade instrument. If future research demonstrates the feasibility of the technique, it is easy to envision the construction of a custom atomic force microscope that could image integrated circuits *in situ* on their circuit boards. A key aspect of the research would be to perform topside imaging and characterization, avoiding backside polishing and etching and, in principle, maintaining the integrity of the chip as a working device.

The prospects for the future are exciting. Micro-electro-mechanical systems technology and other manufacturing processes that produce smaller, lighter atomic force microscopy structures with higher fundamental resonance would allow for increases in data access rates and crash-free investigations of surfaces with high relief. This would affect the practicability of security and forensic applications discussed in this work. Further development of high-speed electronics and microwave engineering may permit other advances in the surface characterization of semiconductor devices, or they may simply make measurements easier, cheaper and more reliable.

From the point of view of forensic investigations, atomic force microscopy has a number of advantages: it is minimally invasive; it may be performed in a wide variety of environments; and it can be used to image almost any kind of sample. The problems in performing atomic force microscopy investigations of working memory chips are mostly practical, but are of sufficient severity to deter casual investigations. Whatever the future may bring in terms of instruments and their capabilities, one thing is clear – sample preparation techniques will always be of the utmost importance.

References

[1] S. Axelsson, A Preliminary Attempt to Apply Detection and Estimation Theory to Intrusion Detection, Technical Report 00-4, Department of Computer Engineering, Chalmers University of Technology, Goteborg, Sweden, 2000.

[2] E. Biham and A. Shamir, Differential fault analysis of secret key cryptosystems, *Proceedings of the Seventeenth Annual International Cryptology Conference*, pp. 513–525, 1997.

[3] G. Binnig, C. Quate and C. Gerber, Atomic force microscope, *Physical Review Letters*, vol. 56, pp. 930–933, 1986.

[4] G. Binnig and H. Rohrer, Scanning tunneling microscopy, *Helvetica Physica Acta*, vol. 55, pp. 726–735, 1982.

[5] Bruker, Innova, Billerica, Massachusetts (`www.bruker.com/pro ducts/surface-and-dimensional-analysis/atomic-force-mic roscopes/innova/overview.html`), 2019.

[6] D. Chiang, P. Lei, F. Zhang and R. Barrowcliff, Dynamic EFM spectroscopy studies on electric force gradients of IrO_2 nanorod arrays, *Nanotechnology*, vol. 16(3), pp. S35–S40, 2005.

[7] C. De Nardi, R. Desplats, P. Perdu, C. Guerin, J. Gauffier and T. Amundsen, Direct measurements of charge in floating gate transistor channels of flash memories using scanning capacitance microscopy, *Proceedings of the Thirty-Second International Symposium on Testing and Failure Analysis*, pp. 86–93, 2006.

[8] S. Garfinkel, Digital forensics research: The next 10 years, *Digital Investigation*, vol. 7(S), S64–S73, 2010.

[9] M. Green, Why Can't Apple Decrypt Your iPhone, *A Few Thoughts on Cryptography Engineering Blog*, October 4, 2014.

[10] J. Halderman, S. Schoen, N. Heninger, W. Clarkson, W. Paul, J. Calandrino, A. Feldman, J. Appelbaum and E. Felten, Lest we remember: Cold-boot attacks on encryption keys, *Communications of the ACM*, vol. 52(5), pp. 91–98, 2009.

[11] J. Kim, D. Son, M. Lee, C. Song, J. Song, J. Koo and D. Kim, A wearable multiplexed silicon nonvolatile memory array using nanocrystal charge confinement, *Science Advances*, vol. 2(1), article no. e1501101, 2016.

[12] S. Kinney, *Trusted Platform Module Basics: Using TPM in Embedded Systems*, Newnes, Burlington, Massachusetts, 2006.

[13] P. Kocher, J. Jaffe and B. Jun, Differential power analysis, *Proceedings of the Nineteenth Annual International Cryptology Conference*, pp. 388–397, 1999.

[14] D. Konopinski, Forensic Applications of Atomic Force Microscopy, Doctoral Dissertation, Department of Electronic and Electrical Engineering, University College London, London, United Kingdom, 2013.

[15] K. Lai, W. Kundhikanjana, H. Peng, Y. Cui, M. Kelly and Z. Shen, Tapping mode microwave impedance microscopy, *Review of Scientific Instruments*, vol. 80(4), 043707, 2009.

[16] M. Maroufi, A. Fowler, A. Bazaei and S. Moheimani, High-stroke silicon-on-insulator MEMS nanopositioner: Control design for non-raster scan atomic force microscopy, *Review of Scientific Instruments*, vol. 86(2), 023705, 2015.

[17] I. Mayergoyz and C. Tse, *Spin-Stand Microscopy of Hard Disk Data*, Elsevier, Oxford, United Kingdom, 2007.

[18] J. Pineda, G. Pascual, B. Kim and K. Lee, Electrical Characterization of Semiconductor Device Using SCM and SKPM Imaging, Application Note #8, Park Systems, Santa Clara, California, 2017.

[19] M. Ranum, Thinking about firewalls, *Proceedings of the Second International Conference on Systems and Network Security and Management*, 1993.

[20] J. Rushby, A trusted computing base for embedded systems, *Proceedings of the Seventh Department of Defense/National Bureau of Standards Computer Security Conference*, pp. 294–311, 1984.

[21] S. Skorobogatov and R. Anderson, Optical fault induction attacks, *Proceedings of the Fourth International Workshop on Cryptographic Hardware and Embedded Systems*, pp. 2–12, 2002.

[22] W. Stallings, *Cryptography and Network Security: Principles and Practice*, Prentice Hall, Upper Saddle River, New Jersey, 2010.

[23] D. von Oheimb, Information flow control revisited: Noninfluence = noninterference + nonleakage, *Proceedings of the Ninth European Symposium on Research in Computer Security*, pp. 225–243, 2004.

[24] A. Waksman and S. Sethumadhavan, Tamper evident microprocessors, *Proceedings of the IEEE Symposium on Security and Privacy*, pp. 173–188, 2010.

Chapter 13

TIMELINE VISUALIZATION OF KEYWORDS

Wynand van Staden

Abstract   Visualizations of communications between actors are typically presented
as actor interactions or as plots of the dates and times when the communications occurred. These visualizations are valuable to forensic analysts; however, they do not provide an understanding of the general flow
of the discussed topics, which are identified by keywords or keyphrases.
The ability to view the content of a corpus as a timeline of discussion
topics can provide clues to when certain topics became more prevalent in
the discussion, when topics disappeared from the discussion and which
topics are outliers in the corpus. This, in turn, may help discover related
topics and times that can be used as clues in further analyses. The goal
is to provide a forensic analyst with assistance in systematically reviewing data, eliminating the need to manually examine large amounts of
communications.

This chapter focuses on the timeline-based visualization of keywords
in a text corpus. The proposed technique employs automated keyword
extraction and clustering to produce a visual summary of topics recorded
from the content of an email corpus. Topics are regarded as keywords
and are placed on a timeline for visual inspection. Links are placed
between topics as the timeline progresses. Placing topics on a timeline
makes it easier to discover patterns of communication about specific
topics instead of merely focusing on general discussion patterns. The
technique complements existing visualization techniques by enabling a
forensic analyst to concentrate on the most interesting portions of a
corpus.

Keywords: Email corpus, topic extraction, timeline visualization

1.   Introduction

Data visualization in social network analysis is typically organized
around communication patterns – who knows whom and how the ac-

© IFIP International Federation for Information Processing 2019
Published by Springer Nature Switzerland AG 2019
G. Peterson and S. Shenoi (Eds.): Advances in Digital Forensics XV, IFIP AICT 569, pp. 239–252, 2019.
https://doi.org/10.1007/978-3-030-28752-8_13

tors interact. During an investigation, social network analysis is used with great effect to determine which actors might be colluding, where to search for potential material evidence and which actors should be interviewed for additional data. When a large corpus has to be examined, the visualization of interactions and communication patterns is an important part of the analysis process.

As discussed in the next section on related work, considerable research has concentrated on data visualization. However, in nearly all cases, visualization has focused on the quantitative aspects of the data for example, plotting email messages as two-dimensional data where one dimension corresponds to the email user and the other dimension corresponds to the time when the email was sent. Such visualization may directly assist in discovering evidence.

Taking a cue from information cartography, this chapter proposes a technique for visualizing topical communication on a timeline by extracting keywords (representative of topics) from a corpus. This provides a means for a forensic analyst to follow the progression of topics (and related topics) based on a particular time window. The idea of providing a timeline visualization of topics during analysis comes from the work of Shahaf et al. [20]. However, the proposed approach is different because, as a first-step technique that assists a forensic analyst in an investigation, it does not hide information and important clues from the analyst. Specifically, it enables a forensic analyst to identify the various topics in a corpus and how they flow through time, providing valuable assistance in understanding actor interactions and supporting focused explorations of a corpus. The timeline-based visualization of keywords is tested on the well-known Enron email corpus.

2.   Related Work

Digital forensics is becoming a mature discipline [13] with standardized processes and commercial tools. Email forensics has developed into a separate area within the discipline and comes with its own challenges [8].

Several researchers have investigated visualization as a means to discover digital evidence. Schrenk and Poisel [18] survey visualization techniques used in digital forensics. Olsson and Boldt [12] have developed a visualization tool that provides timelines for event reconstruction from files and file content. Fei et al. [3] have employed self-organizing maps to discover anomalies in data that can help identify sources of evidence.

Devendran et al. [2] have conducted a comparative study of five popular open-source email forensic tools; their study indicates that visualization is restricted to standard content inspection. However, tools that support email visualization are becoming more prevalent. For example, Stadlinger and Dewald [22] have developed a tool that depicts email communications between different accounts. The tool provides histograms of email volumes per hour and per day, and the most active users. It also creates a link graph that highlights the flow of email from accounts to other accounts (i.e., outbound communication patterns between accounts).

Haggerty et al. [5] report that most investigative tools for email visualization support quantitative data analyses. Their triage system makes use of link analysis and tag clouds to visualize interactions and actor relationships. The tag clouds highlight important words and concepts shared by actors. However, their approach focuses the attention of a forensic analyst on searching for evidence instead of appreciating the patterns and events that are latent in the data. Understanding the nascent patterns and events assists the analyst in developing complex and concrete ideas about where to search for evidence.

Frau et al. [4] have developed a tool that depicts email as glyphs whose size and color change based on their locations in the email folder hierarchy and their overall size. Email messages are presented as a scatter-plot on a timeline.

Viegas et al. [24] have created a forensic tool that provides an overview of topics discussed between users and their contacts. However, their tool is not designed for email visualization and does not link topics over time.

Nordbo [11] has developed a visualization tool that considers user interactivity to discover digital evidence. The tool provides email timeline views per user, activity histograms summarized per day, week and overall, frequencies of messages sent and received, and popular communication times. Several other visualization techniques and tools have been developed and interested readers are referred to the work of Joorabi [7], Appan et al. [1], and Sudarsky and Hjelsvold [23].

Very little research has focused on topics and timelines as a means for visualization. One exception is the work of Shahaf et al. [20], which introduces the concept of information maps ("metro maps") that convey the knowledge and evolution of stories in curated news articles. The maps are generated based on properties and constraints – coherence, coverage and connectivity. A highly-coherent map provides storylines in which each point in the story relates to the previous and next "stops." High coverage ensures that a storyline provides as much information as possible about a story and promotes diversity (i.e., the storyline provides as much information about a particular topic as possible). Connectivity ensures that links between different aspects of a story are provided, meaning that the connections between different aspects of the story are present (as a reader might expect). The concept of metro maps has been extended to the visualization of academic (research) papers [19].

The work of Shahaf et al. [20] has motivated the timeline visualization of topics described in this chapter. However, the notions of coverage and diversity are difficult to apply to an email corpus during an investigation because the objective of a forensic analyst is not to acquire new knowledge. Instead, the analyst is interested in discovering material evidence and may not care how well an email message covers a particular topic, just that the topic is present in the message. The considerations of coverage and diversity in the case of an email corpus and topic visualization are left for future research.

3. Proposed Technique

The proposed technique incorporates three processes: (i) data acquisition; (ii) topic extraction and preprocessing; and (iii) visualization:

- **Data Acquisition:** The data acquisition process involves data preparation, extraction and storage in the appropriate formats and locations. This requires the email messages to be parsed in various formats, such as UNIX mbox [6] and Microsoft PST/OST. Data may be stored in a normalized relational database, in a NoSQL database that handles large data volumes more effectively and scalably, or in a container format. The principle is that it should be easy to query the data.

- **Topic Extraction and Preprocessing:** Topic extraction and preprocessing involve the following steps:

  - Automated extraction of keywords using established techniques such as word co-location analysis and named entity recognition.

  - Normalization of extracted keywords, which includes automated spelling correction. This step can be difficult because most spelling corrections are curated: the user is present and can provide guidance to the spelling corrector. Extracting keywords and making automatic corrections require assumptions to be made about the correct spellings of words, which could form a vernacular that is unique to the entity. This issue is discussed by Samanta and Chaudhuri [17].

  - Generation of common lexicographic rendering indexes related to normalization. The lexicographic renderings of keywords may differ slightly in the corpus due to the writing habits of individuals (e.g., misspellings of words and uncommon renderings of company names). These minor variations are united to present a consistent view of keywords.

- **Visualization:** Visualization involves the following steps:

  - Acceptance of a search query.
  - Finding related or similar keywords.
  - Clustering topics and email messages based on keywords.
  - Rendering topics and keywords on a timeline.

The proposed technique is implemented using a lightweight SQL relational database system to store the data. Email messages are stored in one table. Another table contains the keywords. A bridging table is used to present the many-to-many relationships between email messages and keywords. A final table contains the normalized keywords and their one-to-many relationships.

The reference implementation was evaluated using the well-known Enron dataset. However, timeline comparisons of reported events and corpus events were not performed.

The next two sections describe the topic extraction and preprocessing phase and the visualization phase in detail.

4. Topic Extraction and Preprocessing

The topic extraction and preprocessing phase involves harvesting keywords (keyphrases) from a corpus and preparing the extracted topics for querying. Three types of models may be employed for keyword extraction: (i) statistical models; (ii) supervised models; and (iii) unsupervised models [21]. Each model type has its own advantages and disadvantages for use in different domains [21]. However, a system used by a forensic analyst should provide as much information as possible with little configuration overhead. Specifically, it should provide a starting point for the analyst.

In the case of the reference implementation, an unsupervised model was considered that would be relatively fast and would not require the number of topics to be predetermined (e.g., latent Dirichlet allocation). For this reason, the rapid automated keyword extraction (RAKE) model [15] was selected. Note that the idea was to create a reference implementation that would permit the replacement of one model with another to adapt topic extraction to a particular domain. However, it should be clear that any other model could be used after sufficient testing in a given domain.

Rapid automated keyword extraction considers stop words as boundaries between potential keywords and scores the individual words based on co-occurrence. The highest scoring candidates are used as keywords for the text. The choice of the stop-word list can play a role in the coherence of the keywords that are harvested. Standard stop-word lists were selected for the reference implementation. However, domain-specific stop words may yield additional benefits [9]; this topic is the subject of future research.

The standard stop-word list provided by NLTK [10] introduced too much noise during initial testing. Therefore, the SMART stop-word list [16] was chosen.

The second part of preprocessing involves getting the keywords ready for searching. Stemming was deemed to be an appropriate preprocessing step for information retrieval.

Stemming maps a known word to a common form. Stemming words that have similar lexical and semantic representations produces a common lexicographic representation that can be used to perform string comparisons more easily. This enables an analyst to enter variations of the words to compare against the corpus. For example, stemming "activities" and "activity" maps both words to the same common lexicographic representation.

This work opted for the well-known stemming technique of Porter [14], which maps "activities" and "activity" to "activ." Thus, a forensic analyst could use the search term "activities," but still obtain all the email messages that contain the term "activity."

In order to facilitate searches based on keywords, all the words in the extracted topics were stemmed and an index was created on the actual topics to which the words referred. Words in the search terms were also stemmed and matches were performed on the stemmed words.

A common representation index was created to account for slight variations of keywords such as "securities exchange commission" and "security exchange commission." The common representation index employed a maximum likelihood estimator to produce a consistent visual rendering of such topics. The estimator mapped keywords (using stemming) to the most common representation found in the list of keywords. In this case:

$$c(k) = \text{argmax}_x \delta(x, k)$$

where $\delta(x, k)$ is the number of times x appears as a topic related to k. This approach ensures that minor variations in topic spellings produce a single consistent representation during visualization.

5.   Visualization

The visualization of topics is presented after a search query (term) is issued. The stemming of the search terms works in the same way as the stemming of the keywords obtained via the rapid automated keyword extraction technique. Each keyword is simply stemmed and the stemmed version is used when finding candidate keywords.

5.1   Finding and Ranking

Finding related topics is similar to finding related documents in information retrieval. Several techniques can be used to find related documents, the most common being TF-IDF (term frequency/inverse document frequency). In basic information retrieval, searching and matching are variations of the bag-of-words approach.

Upon conducting sampling and statistical analyses of the words in a collection of documents, it is possible to determine how different the documents are from each other. In such cases, the search term is considered to be a document and the searching system simply finds all the documents that are similar to the search term document. These search techniques are extremely powerful for large corpora with large documents. However, since the approach presented here preprocesses the keywords, the search essentially matches words in the short search term against words in the short keyword/topic list. This makes the matching technique simple and fast.

The matching technique employs a similarity score based on the Jaccard index J, which is defined as:

$$J(x, y) = \frac{|x \cap y|}{|x| + |y|}$$

where x and y are the words being compared.

Using the stemmed index created in the preprocessing stage, all the candidate keywords are found and then ranked according to the Jaccard distance measure. Algorithm 1 specifies the details of the search.

The search results return the top $n = 30$ exact keyword matches in order to reduce the amount of noise that can bleed into the listed keywords and reduce the information presented in the visualization. Note that the parameter n can be adjusted to de-clutter the results.

The final step in the process is to retrieve all the email messages in the corpus that contain the listed keywords. These candidate email messages are used to construct the timeline.

Algorithm 1: Search algorithm.

Input: T: Search term
Result: R: Ranked list of keywords
S ← StemmedWordList(T);
foreach *s in S* **do**
    | C ← C ∪ KeywordLookup(s);
end
foreach *c in C* **do**
    | R ← R ∪ (c,J(c,T))
end
return R;

5.2   Clustering

The results must be clustered prior to visualization. The choice of clustering method impacts the eventual summary and display of data. However, in the case of the reference implementation, it was decided to employ as much information as possible in the clustering. Topics were clustered per day and subsequently by merging related keywords common to email messages on the same day. The resulting cluster contained common keywords in a collection of email messages on a particular day.

To illustrate the clustering technique, assume that the search term "accounting irregularities" yields several email messages on a single day. This gives rise to one of two scenarios: (i) several email messages on the given day, where all the messages contain the same keyword used in the search; or (ii) several email messages on the given day, where some messages contain only the keyword related to the search term and some messages contain the keyword as well as additional keywords.

In the first scenario, clustering emails containing the same keyword would de-clutter a visual rendering of the timeline. In the second scenario, clustering email messages with the same keyword would also de-clutter the display. However, these keywords are not assimilated into multi-keyword clusters because the assimilation could obscure unique email clusters.

5.3   Rendering

The timeline is rendered by displaying topic clusters using the topics in the email messages per day. Each topic cluster is then linked to the cluster corresponding to a following day based entirely on the topic. Linking is performed using the following rules:

- For each topic on a given day, find the first future occurrence of the same topic and create a link to the topic.

- For each topic on a given day, find the first future occurrence of the same topic that is present in a multi-topic cluster and create a link to the topic.

- For each topic in a multi-topic cluster, find the first future occurrence of the topic and create a link to the topic.

6.   Results

This section presents the visualization renderings for searches using two keyphrases: (i) "accounting irregularities;" and (ii) "securities exchange commission." The renderings illustrate the utility of timeline visualization of keywords/keyphrases. Presenting visual renderings on paper is always difficult and the renderings have been reduced in size for presentation purposes.

In the case of the keyphrase "accounting irregularities," some of the top search terms returned – "accounting irregularities," "accounting irregularities disclosed," "accounting irregularities leaked," "creative accounting" – were found in nearly 400 email messages.

Figure 1 shows the visualization corresponding to the keyphrase "accounting irregularities." The timeline clearly shows two interesting periods during which several clusters of email messages discussed accounting irregularities – between January 14 and 18, as well as between January 29 and March 3. Each period contains a flurry of email messages that discuss the same topics. A merge/split on the topics reveals that the phrase "creative accounting" appears in the second period. However, the keyphrase meanders throughout the timeline. The appearances of "creative accounting" and "accounting problem" yield a potential area of interest because it appears that these topics were being discussed after the initial shock of the Enron exposé.

In the case of the keyphrase "securities exchange commission," some of the top search terms returned – "securities exchange commission inquiry," "exchange commission," "exchange commission opens" – were found in nearly 1,700 email messages

Figure 2 shows the visualization corresponding to the keyphrase "securities exchange commission." Since Enron had regular dealings with the U.S. Securities and Exchange Commission (SEC), the keyphrase search does not provide a significant amount of information – the timeline is riddled with references on an almost daily basis. Moreover, since the corpus was collected around the time of the investigation by the U.S. Securities and Exchange Commission, many email messages would be expected to include the keyphrase "securities exchange commission." The point is that a search using this particular keyphrase delivers very few outliers.

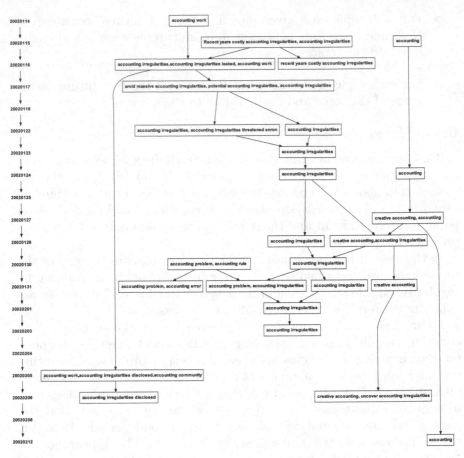

Figure 1. Visualization corresponding to "accounting irregularities."

7.   Conclusions

Most digital forensic investigations engage typical social network visualization approaches that depict person-person communications on a timeline and person-person links. Some approaches even provide keyword listings per day or word clouds. However, these visualizations do not provide an understanding of the general flow of the discussed topics, which are identified by keywords or keyphrases.

The proposed timeline-based visualization of keywords draws on the concept of metro maps of science [19]. It leverages automated keyword extraction and clustering to produce a visual summary of topics in an email corpus. Topics are regarded as keywords and are placed on a timeline for visual inspection; links are then placed between topics as

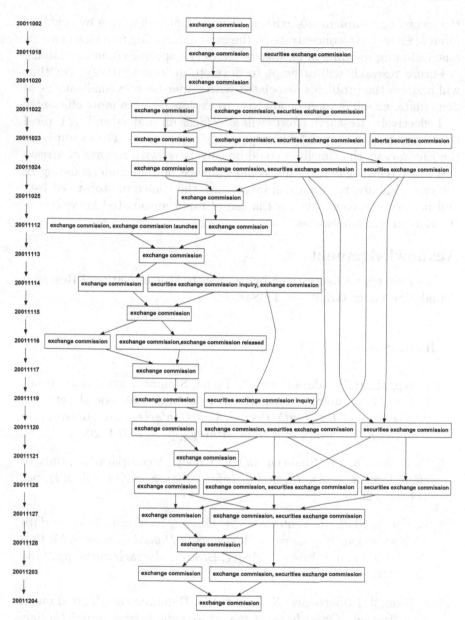

Figure 2. Visualization corresponding to "securities exchange commission."

the timeline progresses. Placing topics on a timeline makes it easier for forensic analysts to discover patterns of communication about specific topics instead of manually analyzing general discussion patterns. Also, the technique complements existing visualization techniques by enabling forensic analysts to concentrate on the most interesting portions of a corpus, including zooming in on specific times and specific communications.

Future research will attempt to develop an interactive system that will address the problems associated with paper-based visualizations of dense data, enabling forensic analysts to explore corpora more efficiently and effectively. Research efforts will also focus on understanding topical conversations in a corpus by incorporating news events. For example, in the Enron case, the timelines could be correlated with reports of arrests and other relevant events that could enhance human understanding of the case. Finally, research will investigate the choice of stop-word lists and perform rigorous tests on the use of rapid automated keyword extraction on email messages.

Acknowledgement

This research was supported by the South African National Research Foundation under Grant No. 114848.

References

[1] P. Appan, H. Sundaram and B. Tseng, Summarization and visualization of communication patterns in a large-scale social network, *Proceedings of the Tenth Pacific-Asia Conference on Advances in Knowledge Discovery and Data Mining*, pp. 371–379, 2006.

[2] V. Devendran, H. Shahriar and V. Clincy, A comparative study of email forensic tools, *Journal of Information Security*, vol. 6(2), pp. 111–117, 2015.

[3] B. Fei, J. Eloff, H. Venter and M. Olivier, Exploring forensic data with self-organizing maps, in *Advances in Digital Forensics*, M. Pollitt and S. Shenoi (Eds.), Springer, Boston, Massachusetts, pp. 113–123, 2005.

[4] S. Frau, J. Roberts and N. Boukhelifa, Dynamic coordinated email visualization, *Proceedings of the Thirteenth International Conference in Central Europe on Computer Graphics, Visualization and Computer Vision*, pp. 187–193, 2005.

[5] J. Haggerty, S. Haggerty and M. Taylor, Forensic triage of email network narratives through visualization, *Information Management and Computer Security*, vol. 22(4), pp. 358–370, 2014.

[6] E. Hall, The application/mbox Media Type, RFC 4155 (`data tracker.ietf.org/doc/rfc4155`), 2005.

[7] M. Joorabchi, EmailTime: Visualization and Analysis of Email Dataset, Master's Thesis, School of Interactive Art and Technology, Simon Fraser University, Burnaby, Canada, 2010.

[8] H. Lalla and S. Flowerday, Towards a standardized digital forensic process, *Proceedings of the Information Security South Africa Conference*, 2010.

[9] M. Makrehchi and M. Kamel, Extracting domain-specific stop words for text classifiers, *Intelligent Data Analysis*, vol. 21(1), pp. 39–62, 2017.

[10] NLTK Project, Natural Language Toolkit (`www.nltk.org`), 2019.

[11] A. Nordbo, Data Visualization for Discovery of Digital Evidence in Email, Master's Thesis, Department of Computer Science and Media Technology, Gjovik University College, Gjovik, Norway, 2014.

[12] J. Olsson and M. Boldt, Computer forensic timeline visualization tool, *Digital Investigation*, vol. 6(S), pp. S78–S87, 2009.

[13] G. Palmer, A Road Map for Digital Forensic Research, DFRWS Technical Report, Technical Report DTR-T001-01 Final, Air Force Research Laboratory, Rome, New York, 2001.

[14] M. Porter, An algorithm for suffix stripping, in *Readings in Information Retrieval*, K. Sparck-Jones and P. Willet (Eds.), Morgan Kaufmann, San Francisco, California, pp. 313–316, 1997.

[15] S. Rose, D. Engel, N. Cramer and W. Cowley, Automatic keyword extraction from individual documents, in *Text Mining: Applications and Theory*, M. Berry and J. Kogan (Eds.), John Wiley and Sons, Hoboken, New Jersey, pp. 1–20, 2010.

[16] G. Salton, *The SMART Retrieval System: Experiments in Automatic Document Processing*, Prentice-Hall, Upper Saddle River, New Jersey, 1971.

[17] P. Samanta and B. Chaudhuri, A simple real-word error detection and correction using local word bigram and trigram, *Proceedings of the Twenty-Fifth Conference on Computational Linguistics and Speech Processing*, pp. 211–220, 2013.

[18] G. Schrenk and R. Poisel, A discussion of visualization techniques for the analysis of digital evidence, *Proceedings of the Sixth International Conference on Availability, Reliability and Security*, pp. 758–763, 2011.

[19] D. Shahaf, C. Guestrin and E. Horvitz, Metro maps of science, *Proceedings of the Eighteenth ACM SIGKDD International Conference on Knowledge Discovery and Data Mining*, pp. 1122–1130, 2012.

[20] D. Shahaf, C. Guestrin and E. Horvitz, Trains of thought: Generating information maps, *Proceedings of the Twenty-First International Conference on World Wide Web*, pp. 899–908, 2012.

[21] S. Siddiqi and A. Sharan, Keyword and keyphrase extraction techniques: A literature review, *International Journal of Computer Applications*, vol. 109(2), pp. 18–23, 2015.

[22] J. Stadlinger and A. Dewald, A forensic email analysis tool using dynamic visualization, *Journal of Digital Forensics, Security and Law*, vol. 12(1), article no. 6, 2017.

[23] S. Sudarsky and R. Hjelsvold, Visualizing electronic mail, *Proceedings of the Sixth International Conference on Information Visualization*, pp. 3–9, 2002.

[24] F. Viegas, S. Golder and J. Donath, Visualizing email content: Portraying relationships from conversational histories, *Proceedings of the SIGCHI Conference on Human Factors in Computing Systems*, pp. 979–988, 2006.

Chapter 14

DETERMINING THE FORENSIC DATA REQUIREMENTS FOR INVESTIGATING HYPERVISOR ATTACKS

Changwei Liu, Anoop Singhal, Ramaswamy Chandramouli and Duminda Wijesekera

Abstract   Hardware/server virtualization is commonly employed in cloud computing to enable ubiquitous access to shared system resources and provide sophisticated services. The virtualization is typically performed by a hypervisor, which provides mechanisms that abstract hardware and system resources from the operating system. However, hypervisors are complex software systems with many vulnerabilities. This chapter analyzes recently-discovered vulnerabilities associated with the Xen and KVM open-source hypervisors, and develops their attack profiles in terms of hypervisor functionality (attack vectors), attack types and attack sources. Based on the large number of vulnerabilities related to hypervisor functionality, two sample attacks leveraging key attack vectors are investigated. The investigation clarifies the evidence coverage for detecting attacks and the missing evidence needed to reconstruct attacks.

Keywords: Cloud computing, hypervisors, Xen, KVM, vulnerabilities, forensics

1.   Introduction

Most cloud services are provided by virtualized environments. Virtualization is a key feature of cloud computing that enables ubiquitous access to shared pools of system resources and high-level services provisioned with minimal management effort [15, 28]. Although an operating system directly controls hardware resources, virtualization via a hypervisor or virtual machine monitor (VMM) [6] in a cloud environment provides an abstraction of hardware and system resources. The hypervisor serves as a software layer between the physical hardware and virtual (or

© IFIP International Federation for Information Processing 2019
Published by Springer Nature Switzerland AG 2019
G. Peterson and S. Shenoi (Eds.): Advances in Digital Forensics XV, IFIP AICT 569, pp. 253–272, 2019.
https://doi.org/10.1007/978-3-030-28752-8_14

guest) machines, presenting the guest operating systems with virtual operating platforms and managing their execution. However, hypervisors are complex software systems with numerous lines of code and many vulnerabilities [20]. These vulnerabilities can be exploited to gain access to and control a hypervisor, and subsequently attack all the virtual machines that execute in the compromised hypervisor.

Several researchers have characterized and assessed hypervisor vulnerabilities, developed tools for detecting vulnerabilities and identified evidence that can be used for attack forensics [7, 9, 19, 20, 26, 27]. Hypervisor forensics seeks to extract leftover artifacts in order to investigate and analyze attacks at the hypervisor level. Techniques such as inspecting physical memory to locate evidence of attacks have been explored [7], but little, if any, research has analyzed recent hypervisor vulnerabilities to derive attack profiles and leverage them to discover forensic evidence that can help reconstruct hypervisor attacks.

The research described in this chapter was motivated by the work of Perez-Botero et al. [20] that characterized hypervisor vulnerabilities with the objective of preventing their exploitation. This chapter focuses on recent (2016 and 2017) vulnerability reports associated with the popular, open-source Xen and KVM hypervisors, which are listed in the National Institute of Standards and Technology National Vulnerability Database (NIST-NVD). The chapter analyzes and classifies the vulnerabilities to derive attack profiles based on hypervisor functionality, attack type and source. Additionally, it simulates sample attacks to ascertain their forensic data coverage and explore methods for identifying and obtaining the evidence needed to reconstruct attacks.

2.   Background and Related Work

This section discusses hypervisor architectures with a focus on the Xen and KVM hypervisors. Also, it discusses related work in the area of cloud forensics.

2.1   Hypervisors

Hypervisors are software and/or firmware modules that virtualize system resources such as CPU, memory and devices. Popek and Goldberg [21] have classified hypervisors as Type 1 and Type 2 hypervisors. A Type 1 hypervisor runs directly on the host hardware to control the hardware and manage guest operating systems. For this reason, a Type 1 hypervisor is sometimes called a "bare metal" hypervisor. Example Type 1 hypervisors are Xen, Microsoft Hyper-V and VMware ESX/ESXi.

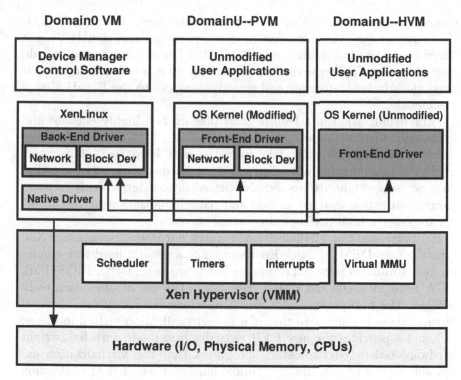

Figure 1. Xen hypervisor architecture.

A Type 2 hypervisor is similar to a program that executes as a process in an operating system. Example Type 2 hypervisors are VMware Player, VirtualBox, Parallels Desktop for Mac and QEMU.

Some hypervisors have features of Type 1 and Type 2 hypervisors. For example, the Linux-kernel-based virtual machine (KVM) is a kernel module that effectively converts the host operating system to a Type 1 hypervisor, but it is also categorized as a Type 2 hypervisor because a Linux distribution is a general-purpose operating system that executes other applications that compete for virtual machine resources [18].

A market report [14] lists the most popular hypervisors as Microsoft Hyper-V, VMware VSphere/ESX, Citrix XenServer/Xen and KVM. Because Microsoft Hyper-V and VMware VSphere/ESX are commercial products, this work focuses on the remaining two hypervisors, Xen and KVM, which are both open source.

Xen Hypervisor. Figure 1 shows the Xen hypervisor architecture. The hypervisor manages three kinds of virtual machines. The first is the control domain (Dom0) and the other two are guest domains (DomU) that support two virtualization modes – paravirtualization (PV) and hardware-assisted virtualization (HVM) [31]. Dom0 is the initial domain started by the Xen hypervisor upon booting up a privileged domain. It plays the administrator role and provides services to the DomU virtual machines.

Paravirtualization in a DomU guest domain is a highly efficient and lightweight virtualization technology introduced by Xen that does not require virtualization extensions from the host hardware. Thus, paravirtualization enables virtualization on a hardware architecture that does not support hardware-assisted virtualization; however, it requires paravirtualization-enabled kernels and paravirtualization drivers in order to power a high performance virtual server.

Hardware-assisted virtualization requires hardware extensions. Xen typically uses QEMU (Quick Emulator) [22], a generic hardware emulator that simulates personal computer hardware (e.g., CPU, BIOS, IDE, VGA, network cards and USBs). Because of the use of simulation technologies, the performance of a virtual machine with hardware-assisted virtualization is inferior to that of a paravirtualization virtual machine.

Xen 4.4 provides the new PVH virtualization mode with lightweight hardware-assisted-virtualization-like guests that use virtualization extensions in the host hardware. Unlike hardware-assisted virtualization guests that (for example) use QEMU to emulate devices, PVH guests use paravirtualization drivers for input/output and native operating system interfaces for virtualized timers, virtualized interrupts and booting. A PVH guest also requires a PVH-enabled guest operating system [31].

KVM Hypervisor. KVM was introduced in 2006 and it was soon merged into the Linux kernel (2.6.20) in open-source hypervisor projects. KVM is a full virtualization solution for Linux that runs on x86 hardware with virtualization extensions (Intel VT or AMD-V) to enable virtual machines to execute as normal Linux processes [11].

Figure 2 shows a KVM hypervisor architecture that uses QEMU to create guest virtual machines that execute as separate user processes. KVM is considered to be a Type 2 hypervisor because it is installed on top of the host operating system. However, the KVM kernel module turns the Linux kernel into a Type 1 bare-metal hypervisor, providing the functionality of the most complex and powerful Type 1 hypervisors.

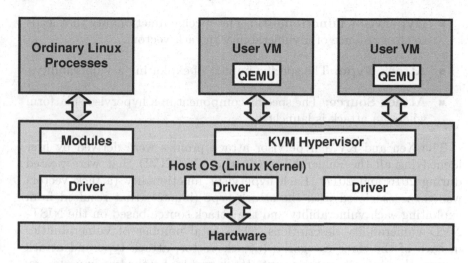

Figure 2. KVM hypervisor architecture.

2.2   Related Work

Hypervisor attacks are defined as exploits of hypervisor vulnerabilities that enable external attackers to gain access to and control hypervisors [24]. Perez-Botero et al. [20] have characterized Xen and KVM vulnerabilities based on hypervisor functionality. However, the characterizations cannot be used to predict attack trends.

Thongthua and Ngamsuriyaroj [27] have assessed the vulnerabilities of popular hypervisors, including VMware ESXi, Citrix XenServer and KVM, using the NIST 800-115 security testing framework. Their assessments of vulnerabilities cover weaknesses, severity scores and attack impacts.

Joshi et al. [9] have researched threats to hypervisors and hypervisor forensic mechanisms. They also discuss the use of virtual machine introspection at the hypervisor level to detect attacks [5, 19] and memory forensic techniques to identify hypervisor attack artifacts in a host memory dump [7].

3.   Deriving Hypervisor Attack Profiles

As a prelude to determining the forensic data requirements for detecting hypervisor attacks, the following criteria are employed to derive attack profiles based on recent hypervisor vulnerabilities:

- **Hypervisor Functionality:** The specific functionality that leads to the existence of a vulnerability (attack vector).

- **Attack Type:** The specific impact of exploiting a vulnerability.

- **Attack Source:** The specific component in a hypervisor platform where an attack is launched.

The Xen and KVM hypervisor attack profiles were derived by first identifying all the vulnerabilities in the NIST-NVD that were posted during 2016 and 2017. Each hypervisor functionality (attack vector) was then associated with the attack type (impact) that resulted from exploiting each vulnerability and the attack source based on the NIST-NVD vulnerability descriptions. The total number of vulnerabilities in each of the three categories (attack vector, attack type and attack source) thus constitute the recent attack profiles for the two hypervisors.

3.1   NIST-NVD Vulnerabilities

The NIST-NVD is a repository of standards-based vulnerability management data, which includes databases of security checklist references, security-related software flaws, misconfigurations, product names and impact metrics [17]. A search of the NIST-NVD for vulnerabilities posted in 2016 and 2017 revealed 83 Xen hypervisor vulnerabilities and 20 KVM hypervisor vulnerabilities. These vulnerabilities were associated with the three classification criteria: (i) hypervisor functionality; (ii) attack type; and (iii) attack source.

3.2   Hypervisor Functionality

In order to better understand hypervisor vulnerabilities, Perez-Botero et al. [20] considered eleven traditional hypervisor functionalities and mapped vulnerabilities to them. The eleven vulnerabilities are: (i) virtual CPUs; (ii) virtual symmetric multiprocessing; (iii) soft memory management units; (iv) input/output and networking; (v) paravirtualized input/output; (vi) interrupt and timer mechanisms; (vii) hypercalls; (viii) VMExits; (ix) virtual machine management; (x) remote management software; and (xi) hypervisor add-ons. Based on the common functions provided by functionalities (iv) and (v), the two were merged as a single functionality.

- **Virtual CPUs:** A virtual CPU (also called a virtual processor) abstracts a share of a physical CPU assigned to a virtual machine. The hypervisor allocates a portion of the physical CPU cycles to a virtual CPU assigned to a virtual machine. The hypervisor also schedules virtual CPU tasks to the physical CPUs.

- **Virtual Symmetric Multiprocessing:** Virtual symmetric multiprocessing enables multiple virtual CPUs belonging to the same virtual machine to be scheduled on a physical CPU with at least two logical processors.

- **Soft Memory Management Units:** A memory management unit (MMU) is the hardware that manages memory by translating the virtual addresses manipulated by software to physical addresses. In a virtualized environment, the hypervisor emulates the memory management unit – called the soft memory management unit – of a guest operating system by mapping what the guest operating system sees as physical memory (called pseudo-physical/physical addresses in Xen) to the underlying memory of the machine (called machine addresses in Xen). The physical address to machine address mapping table is typically maintained in the hypervisor and hidden from a guest operating system using a shadow page table for the guest virtual machine. Each shadow page table mapping, which translates virtual addresses of programs in a guest virtual machine to guest (pseudo) physical addresses, is placed in the guest operating system [10, 30].

  The Xen paravirtualized memory management unit model requires a guest operating system to be directly aware of the mapping between (pseudo) physical and machine addresses (using a P2M table). Additionally, in order to read page table entries that contain machine addresses and convert them back to (pseudo) physical addresses, translation from machine to (pseudo) physical addresses via a reverse M2P table is required by the Xen paravirtualized memory management unit model [30].

- **Input/Output and Networking:** A hypervisor provides input/output services to guest virtual machines via three common approaches (all of which are employed by Xen): (i) the hypervisor emulates a known input/output device in a fully virtualized system and each guest uses a native unmodified driver to interact with it; (ii) a paravirtual front-end driver in a paravirtualized system is installed in a modified guest operating system in DomU, which uses shared-memory asynchronous buffer-descriptor rings to communicate with the back-end input/output driver in the hypervisor; and (iii) the host assigns a pass-through device directly to the guest virtual machine. Scalable self-virtualizing input/output devices that allow direct access interface to multiple virtual machines are also employed to reduce input/output virtualization overhead and improve virtual machine performance.

Although hypervisors enforce isolation across virtual machines residing in a single physical machine, Xen uses the grant mechanism for inter-domain communications. Shared-memory communications between unprivileged domains are implemented via grant tables [10]. Grant tables protect the input/output buffers in guest domain memory and share the buffers with Dom0, which enable split device drivers with block device and network interface card input/output. Each domain has its own grant table that allows the domain to inform Xen about the permissions that other domains have on their pages.

KVM typically uses Virtio, a virtualization standard for network and disk drivers. Virtio is architecturally similar to Xen paravirtualized device drivers that comprise front-end and back-end drivers.

- **Interrupt and Timer Mechanisms:** Hypervisors should be able to virtualize and manage interrupts and timers [25], interrupt/ timer controllers of guest operating systems and guest operating system accesses to controllers. The interrupt and timer mechanisms in a hypervisor include a programmable interval timer, advanced programmable interrupt controller and interrupt request mechanisms [20].

- **Hypercalls:** Hypercalls are similar to system calls (syscalls) that provide user-space applications with kernel-level operations. They are used like syscalls with up to six arguments passed in registers. A hypercall layer, which is commonly available, enables guest operating systems to make requests to the host operating system. Domains use hypercalls to request privileged operations from hypervisors such as updating page tables. As a result, an attacker can use hypercalls to attack a hypervisor from a guest virtual machine.

- **VMExits:** Belay at el. [1] describe VMExits as changing virtual machines from the non-root mode to the root mode. VMExits are triggered by certain events in guest virtual machines – external interrupts, triple faults, task switches, input/output operation instructions (e.g., INB and OUTB) and control register accesses. VMExits are the main source of performance degradation in virtualized systems.

- **Virtual Machine Management:** Hypervisors support basic virtual machine management functionality, including starting, pausing and stopping virtual machines. They are implemented in Xen's Dom0 and KVM's `libvirt` driver.

- **Remote Management Software:** Remote management software serves as an interface that connects directly to a hypervisor, providing additional management and monitoring tools. By providing an intuitive user interface that visualizes system status, remote management software enables an administrator to provision and manage virtualized environments.

- **Hypervisor Add-Ons:** Hypervisor add-ons use modular designs to add extended functions. By leveraging interactions between add-ons and a hypervisor, an attacker can cause a host to crash (denial-of-service attack) or even compromise the host.

3.3 Deriving Attack Profiles

Based on the descriptions posted in the NIST-NVD, the 83 Xen and 20 KVM vulnerabilities identified during 2016 and 2017 were mapped to the ten hypervisor functionalities. In order to derive the hypervisor attack profiles, the vulnerabilities were analyzed and classified according to functionality, attack type (impact) and attack source.

Table 1 shows the Xen and KVM vulnerabilities classified by functionality. The numbers of vulnerabilities and their percentages are listed for each hypervisor functionality. With the exception of virtual symmetric multiprocessing, all the other functionalities were reported as having vulnerabilities.

The table reveals that Xen has more vulnerabilities than KVM; one reason may be Xen's broader user base. Furthermore, approximately 71% of the vulnerabilities in Xen and 45% of the vulnerabilities in KVM are concentrated in two functionalities – soft memory management units, and input/output and networking. A detailed analysis of CVE reports indicates that these vulnerabilities primarily originate in page tables and input/output grant table emulation.

Additionally, the vulnerabilities based on input/output and networking functionality were associated with each of the four types of input/output virtualization: fully virtualized devices, paravirtualized devices, direct access devices and self-virtualized devices. The associations revealed that most of the input/output and networking vulnerabilities in Xen come from paravirtualized devices and all the input/output and networking vulnerabilities in KVM come from fully-virtualized devices. In the case of Xen, this is because, in most Xen deployments, input/output and networking functionality is configured using a paravirtualized

Table 1.   Xen and KVM vulnerabilities classified by functionality.

No.	Hypervisor Functionality	Xen	KVM
1	Virtual CPUs	6 (7%)	4 (20%)
2	Virtual Symmetric Multiprocessing	0 (0%)	0 (0%)
3	Soft Memory Management Units	34 (41%)	5 (25%)
4	Input/Output and Networking	24[1] (30%)	4[2] (20%)
5	Interrupt and Timer Mechanisms	7 (8%)	3 (15%)
6	Hypercalls	3 (4%)	1 (5%)
7	VMExits	1 (1%)	2 (10%)
8	Virtual Machine Management	7 (8%)	0 (0%)
9	Remote Management Software	1 (1%)	0 (0%)
10	Hypervisor Add-Ons	0 (0%)	1 (5%)

[1]Five are fully-virtualized; 19 are paravirtualized; none are direct access or self-virtualized
[2]All are fully-virtualized

Table 2.   Types of attacks leveraging Xen and KVM vulnerabilities.

Attack Type	Xen	KVM
Denial-of-Service	48[1] (44%)	17[2] (63%)
Privilege Escalation	33[3] (30%)	3[4] (11%)
Information Leakage	15[5] (14%)	5 (19%)
Arbitrary Code Execution	8[6] (7%)	2[7] (7%)
Reading/Modifying/Deleting Files	3 (3%)	0 (0%)
Others (e.g., Host Compromise, Canceling Other Administrators' Operations and Data Corruption)	3 (3%)	0 (0%)

[1]Four have other impacts; [2]Three have other impacts; [3]Sixteen have other impacts
[4]Two have other impacts; [5]Five have other impacts; [6]Two have other impacts
[7]All have other impacts

device. In the case of KVM, the functionality is configured using a fully virtualized device.

Table 3. Numbers of attacks from various attack sources.

Attack Source	Xen	KVM
Cloud Administrators	2^1 (2%)	0 (0%)
Guest Operating System Administrators	17^2 (20%)	1 (5%)
Guest Operating System Users	63^3 (76%)	17^4 (85%)
Remote Users	1 (1%)	1^5 (5%)
Host Operating System Users	0 (0%)	1 (5%)

[1]Management; [2]Including HVM and PV administrators
[3]Including ARM, x86, HVM and PV users
[4]Including KVM L1, L2 and privileged users
[5]Authenticated remote guest users

Table 2 shows the types of attacks that leverage vulnerabilities in Xen and KVM hypervisors. The most common attack type is denial-of-service (44% for Xen and 63% for KVM), indicating that attacking the availability of cloud services could be a serious cloud security problem. The other top attacks are privilege escalation (30% for Xen and 11% for KVM), information leakage (14% for Xen and 19% for KVM) and arbitrary code execution (7% for Xen and 7% for KVM). Although these attacks occur with less frequency than denial-of-service attacks, they result in more serious damage by enabling attackers to obtain sensitive user information or compromise hosts or guest virtual machines.

Table 3 shows the numbers of attacks for various attack sources. The greatest source of attacks is guest operating system users (76% for Xen and 85% for KVM); other attack sources are cloud administrators, guest operating system administrators and remote users. This suggests that cloud providers should monitor guest user activities in order to reduce the risk of attacks.

4. Sample Attacks and Forensic Implications

Because the Xen hypervisor attack profile lists many vulnerabilities related to the soft memory management unit functionality and guest virtual machines as the major attack source, two sample attacks were executed to explore evidence coverage and methodologies for identifying evidence in order to reconstruct hypervisor attacks. One attack exploited the CVE-2017-7228 vulnerability and the other attack exploited the CVE-2016-6258 vulnerability.

4.1   Sample Attacks

As discussed earlier, the Xen hypervisor manages three kinds of virtual machines, the control domain (Dom0) and guest domains (DomU) that support two virtualization modes, paravirtualization and hardware-assisted virtualization. The paravirtualization mode is popular due to its better performance [4]. However, because the Xen paravirtualization model uses complex code to emulate memory management units, vulnerabilities such as CVE-2017-7228 and CVE-2016-6258 are introduced.

The CVE-2017-7228 vulnerability was reported in 2017 [8]. This vulnerability in x86 64-bit Xen (versions 4.8.x, 4.7.x, 4.6.x, 4.5.x and 4.4.x) was caused by inadequate checking of the XENMEM_exchange function, which enables a paravirtualization guest user to access hypervisor memory outside the memory provisioned to the guest virtual machine. Therefore, a malicious 64-bit paravirtualization guest user who makes a hypercall HYPERVISOR_memory_op function invoke the XENMEM_exchange function may be able to access all the system memory, enabling a virtual machine escape from DomU to Dom0 (i.e., breaking out of the guest virtual machine and interacting with the hypervisor host operating system) to cause a hypervisor host crash or information leakage.

The CVE-2016-6258 vulnerability was reported in 2016 [2]. The paravirtualization module uses a page table to map pseudo-physical/physical addresses seen by a guest virtual machine to the underlying memory of the machine. Exploiting a vulnerability in the Xen paravirtualization page tables that enables unauthorized modifications to page table entries, a malicious paravirtualization guest can access the page directory with an updated write privilege and execute a virtual machine escape to break out of DomU and control Dom0.

The two attacks were executed on a paravirtualization module configured in Qubes 3.1 and Debian 8 with Xen 4.6. Figure 3 illustrates the attacks. The attacker impersonates a paravirtualization guest root user (the bottom terminal is the attacker's virtual machine (attacker)). The attacker executes the command qvm-run victim firefox to run the Firefox web browser in the paravirtualization guest user virtual machine (victim). This opens a webpage in the victim's virtual machine (shown in the upper window). Note that the qvm-run command in Qubes 3.1 can only be executed by Dom0 to run an application in a guest virtual machine. As shown in Figure 3, both the attacks enable a paravirtualization guest user to gain control of Dom0.

Figure 3.   CVE-2017-7228 and CVE-2016-6258 attacks.

4.2   Identifying Evidence Coverage

The two attacks from guest virtual machines exploited hypercall and soft memory management unit vulnerabilities in Xen. In addition to using Xen's device activity logs, the runtime syscalls of the impacted processes were logged. Subsequent analysis revealed that the syscalls obtained from the attacker's virtual machine constitute useful evidence.

Figure 4 shows the syscalls from the attacking process in the attacker's virtual machine (some of the syscalls are not presented due to space limitations). Specifically, the figure shows that:

- The attacker executed an attack program with the arguments `qvm-run victim firefox` targeting the victim's guest virtual machine (Line 1).

- The attack program and the required Linux libraries were loaded in memory for program execution (Lines 2–4).

```
1.   execve("./attack", ["./attack", "qvm-run victim firefox"], [/* 30
     vars */]) = 0
2.   brk(NULL)                                      = 0x8cd000
3.   mmap(NULL, 4096, PROT_READ|PROT_WRITE, MAP_PRIVATE|MAP_ANONYMOUS,
     -1, 0) = 0x7fa3a3022000
4.   ...
5.   mprotect(0x7fa3a2df9000, 16384, PROT_READ) = 0
6.   mprotect(0x600000, 4096, PROT_READ) = 0
7.   mprotect(0x7fa3a3023000, 4096, PROT_READ) = 0
8.   ...
9.   open("test.ko", O_RDONLY) = 3
10.  finit_module(3, "user_shellcmd_addr=1407334317317"..., 0) = 0
11.  fstat(1, {st_mode=S_IFCHR|0620, st_rdev=makedev(136, 0), ...})
     = 0
12.  mmap(NULL, 4096, PROT_READ|PROT_WRITE, MAP_PRIVATE|MAP_ANONYMOUS,
     -1, 0) = 0x7fa3a3021000
13.  mmap(0x600000000000, 4096, PROT_READ|PROT_WRITE, MAP_PRIVATE|MAP_
     FIXED|MAP_ANONYMOUS|MAP_LOCKED, -1, 0) = 0x600000000000
14.  delete_module("test", O_NONBLOCK) = 0
15.  exit_group(0)= ?

     execve(): Executes the program pointed to by the first argument
     brk(): Changes the location of the program break, which defines
     the end of the process data segment
     mmap(): Creates a new mapping in the virtual address space of the
     calling process
     mprotect(): Changes the access protections of the calling process
     memory pages
     open(): Opens the file test.ko
     finit\_module(): Loads the kernel module test.ko
     fstat(): Gets the file status; the first argument is the file
     descriptor
     delete\_module(): Unloads the injected module
     exit\_group(): Exits all the process threads
```

Figure 4. Syscalls intercepted from the attacking program.

- The memory pages of the attack program were protected from
  being accessed by other processes (Lines 5–8).

- The attack program injected the test.ko loadable Linux module
  into kernel space to exploit the vulnerability, and subsequently
  deleted the module (Lines 9–15).

Despite the noise in the syscalls (a common occurrence), other syscalls
– such as those in Lines 1, 9 and 15 – reveal that the attack program
injected a loadable kernel module into kernel space and proceeded to ex-

ploit the vulnerability to control Dom0. This opened the Firefox browser
in the victim's guest virtual machine.

Clearly, the device activity logs and runtime syscalls constitute valuable evidence in a forensic investigation. Liu et al. [13] have shown that
such evidence helps reconstruct attack paths in attack scenarios. Specifically, during the reconstruction process, an attack path with missing
attack steps drives the search for and collection of additional supporting
evidence.

Analysis of the syscalls captured during the two sample attacks reveal that, while the syscalls obtained from the attacker's virtual machine
are useful for forensic analysis, they lack attack details. In particular,
the syscalls do not provide details of how features of the loadable kernel
module used Xen's memory management to launch the attacks. Another
deficiency is that the syscalls were collected from the attacker's guest virtual machine, which could have been easily compromised or manipulated
by the attacker. Therefore, it is important to implement the monitoring
of virtual machines from the hypervisor to obtain supporting evidence
that would be admissible in legal proceedings.

4.3   Using Virtual Machine Introspection

Virtual machine introspection can be used to inspect the state of
a virtual machine from a privileged virtual machine or the hypervisor
to analyze the software running on the virtual machine [5]. The state
information includes CPU state (e.g., registers), the entire memory and
all input/output device states (e.g., contents of storage devices and the
registers of input/output controllers).

The following virtual-machine-introspection-based forensic applications are promising:

- A virtual-machine-introspection-based application takes a snapshot of the entire memory and the input/output state of a victim's
  virtual machine. The captured state of the running victim virtual
  machine can be compared against a suspended virtual machine in
  a known good state or against the original virtual machine image
  from which the victim virtual machine was instantiated [5].

- A virtual-machine-introspection-based application analyzes execution paths of the compromised virtual machine by tracing the sequence of virtual machine activities and the corresponding complete virtual machine state (e.g., memory map, input/output access). A detailed attack graph is then constructed with virtual
  machine states as nodes and virtual machine activities as edges,
  helping trace the path leading to the compromised state [16].

```
1.   root@debian:/home/guest/src/libvmi/libvmi# ./examples/vmi-process-
     list pv-attacker
2.   Process listing for VM pv-attcker (id=2)
3.   [    0] swapper/0 (struct addr:ffffffff81e13500)
4.   [    1] systemd (struct addr:ffff88--7c460000)
5.   ...
6.   [  674] (sd-pam) (struct addr:ffff880076104600)
7.   [  677] bash (struct addr:ffff880003c8aa00)
8.   [  703] sudo (struct addr:ffff880004341c00)
9.   [  704] attack (struct addr:ffff880004343800)

10.  root@debian:/home/guest/srv/libvmi/libvmi# ./examples/vmi-module-
     list pv-attacker
11.  test
12.  x86_pkg_temp_thermal
14.  Coretemp
15.  crct10dif_pclmul
16.  ...
```

Figure 5.   Running processes and injected modules in the attacker's virtual machine.

Although virtual machine introspection addresses deficiencies in forensic analyses based on system calls from a compromised virtual machine, virtual machine introspection applications must reconstruct the operational semantics of the guest operating system based on low-level sources such as physical memory and CPU registers [3]. Because LibVMI [12] provides virtual machine introspection functionality on Xen and KVM, and bridges the semantic gap by reconstructing high-level state information from low-level physical memory data, experiments were performed using LibVMI as an introspection tool to capture evidence related to the two sample attacks. This was accomplished by installing Xen 4.6 in Debian 8 with the privileged Dom0 and configuring the two paravirtualization guests in DomU with Kernel 3.10.100 and Ubuntu 16.04.5, respectively. LibVMI (release 0.12) [12] installed on Dom0 was employed to capture all the running processes and the Linux modules injected in the attacker's guest virtual machine.

Figure 5 shows the running processes and injected modules in the attacker's virtual machine during the CVE-2017-7228 attack. Lines 1 and 10 show that two programs, vmi-process-list and vmi-module-list, were executed to capture the running processes and modules in the attacker's virtual machine (pv-attacker). Lines 3–9 are the captured processes (each line lists the process number, process name and kernel task list address where the process name was retrieved). Lines 11 to 16 provide information about the captured modules; each line shows the module name. Comparisons of the captured processes and modules during the attack against those collected at an earlier time help identify the attack process (`attack`) in Line 9 and the injected attack module (`test`) in Line 11. The module file extension `.ko` is omitted by the program.

While an introspection tool such as LibVMI is effective at detecting hypervisor attacks, it has some limitations. First, in order to access memory consistently, the tool pauses and resumes the guest virtual machine – the experiment revealed that LibVMI paused the attacker's virtual machine for 0.035756 seconds and 0.036173 seconds when capturing the running processes and injected modules, respectively. Second, because virtual machine introspection is only effective during an attack, an attacker could easily utilize an in-virtual-machine timing mechanism (e.g., `kprobes`, a tracing framework built into the kernel) to evade passive virtual machine introspection [29]. Third, storing the captured snapshots of guest VMs for forensic analyses often requires large amounts of storage space.

5. Conclusions

The analysis and classification of recently-reported Xen and KVM vulnerabilities have contributed to the creation of hypervisor attack profiles. The profiles reveal that most attacks on the two hypervisors are due to vulnerabilities arising from the soft memory management unit and the input/output and networking functionalities; the two most common types of hypervisor attacks are denial-of-service and privilege escalation; and most attacks originate from guest virtual machines.

Experiments involving two sample attacks on the Xen and KVM hypervisors provide insights into evidence coverage and the evidence needed to reconstruct attacks during forensic investigations. The most valuable evidence resides in runtime system memory, and obtaining this evidence with guaranteed integrity requires virtual machine introspection techniques that examine the states of guest virtual machines from the hypervisor level while ensuring strong isolation from the guest virtual machines.

Future research will focus on constructing detailed attack paths from the snapshots of attackers' virtual machines, and addressing the timing and memory issues that come into play when using virtual machine introspection.

This chapter is not subject to copyright in the United States. Commercial products are identified in order to adequately specify certain procedures. In no case does such an identification imply a recommendation or endorsement by the National Institute of Standards and Technology, nor does it imply that the identified products are necessarily the best available for the purpose.

References

[1] A. Belay, A. Bittau, A. Mashtizadeh, D. Terei, D. Mazieres and C. Kozyrakis, Dune: Safe user-level access to privileged CPU features, *Proceedings of the Tenth USENIX Symposium on Operating Systems Design and Implementation*, pp. 335–348, 2012.

[2] J. Boutoille and G. Campana, Xen Exploitation Part 3: XSA-182, Qubes escape, *Quarkslab's Blog* (blog.quarkslab.com/ xen-exploitation-part-3-xsa-182-qubes-escape.html), August 4, 2016.

[3] B. Dolan-Gavitt, B. Payne and W. Lee, Leveraging Forensic Tools for Virtual Machine Introspection, School of Computer Science, Georgia Institute of Technology, Atlanta, Georgia, 2011.

[4] H. Fayyad-Kazan, L. Perneel and M. Timmerman, Full and paravirtualization with Xen: A performance comparison, *Journal of Emerging Trends in Computing and Information Sciences*, vol. 4(9), pp. 719–727, 2013.

[5] T. Garfinkel and M. Rosenblum, A virtual machine introspection based architecture for intrusion detection, *Proceedings of the Network and Distributed System Security Symposium*, pp. 191–206, 2003.

[6] R. Goldberg, Survey of virtual machine research, *IEEE Computer*, vol. 7(9), pp. 34–45, 1974.

[7] M. Graziano, A. Lanzi and D. Balzarotti, Hypervisor memory forensics, *Proceedings of the Sixteenth International Symposium on Research in Attacks, Intrusions and Defenses*, pp. 21–40, 2013.

[8] J. Horn, Pandavirtualization: Exploiting the Xen Hypervisor, Project Zero, Google, Mountain View, California (googleproject zero.blogspot.com/2017/04/pandavirtualization-exploitin g-xen.html), April 7, 2017.

[9] L. Joshi, M. Kumar and R. Bharti, Understanding threats to hypervisor, its forensics mechanism and its research challenges, *International Journal of Computer Applications*, vol. 119(1), pp. 1–5, 2015.

[10] J. Kloster, J. Kristensen and A. Mejlholm, Efficient Memory Sharing in the Xen Virtual Machine Monitor, DAT5 Semester Thesis Report, Department of Computer Science, Aalborg University, Aalborg, Denmark, 2006.

[11] KVM Contributors, Kernel Virtual Machine, KVM (www.linux-kvm.org/page/Main_Page), 2019.

[12] LibVMI Community, LibVMI: LibVMI Virtual Machine Introspection, LibVMI (libvmi.com), 2019.

[13] C. Liu, A. Singhal and D. Wijesekera, A layered graphical model for cloud forensic mission attack impact analysis, in *Advances in Digital Forensics XIV*, G. Peterson and S. Shenoi (Eds.), Springer, Cham, Switzerland, pp. 263–289, 2018.

[14] S. Lowe, 2015 State of Hyperconverged Infrastructure Market Report, ActualTech Media, Bluffton, South Carolina, 2015.

[15] P. Mell and T. Grance, Sidebar: The NIST definition of cloud computing, *Communications of the ACM*, vol. 53(6), p. 50, 2010.

[16] A. Moser, C. Kruegel and E. Kirda, Exploring multiple execution paths for malware analysis, *Proceedings of the IEEE Symposium on Security and Privacy*, pp. 231–245, 2007.

[17] National Institute of Standards and Technology, NIST National Vulnerability Database, Gaithersburg, Maryland (nvd.nist.gov), 2019.

[18] B. Pariseau, KVM reignites Type 1 vs. Type 2 hypervisor debate, TechTarget, Newton, Massachusetts (searchservervirtualizati on.techtarget.com/news/2240034817/KVM-reignites-Type-1-vs-Type-2-hypervisor-debate), April 15, 2011.

[19] B. Payne, Simplifying Virtual Machine Introspection Using Lib-VMI, Sandia Report SAND2012-7818, Sandia National Laboratories, Albuquerque, New Mexico, 2012.

[20] D. Perez-Botero, J. Szefer and R. Lee, Characterizing hypervisor vulnerabilities in cloud computing servers, *Proceedings of the International Workshop on Security in Cloud Computing*, pp. 3–10, 2013.

[21] G. Popek and R. Goldberg, Formal requirements for virtualizable third generation architectures, *Communications of the ACM*, vol. 17(7), pp. 412–421, 1974.

[22] QEMU, QEMU – The FAST! Processor Emulator (www.qemu.org), 2019.

[23] J. Satran, L. Shalev, M. Ben-Yehuda and Z. Machulsky, Scalable I/O – A well-architected way to do scalable, secure and virtualized I/O, *Proceedings of the Workshop on I/O Virtualization*, 2008.

[24] J. Shi, Y. Yang and C. Tang, Hardware assisted hypervisor introspection, *SpringerPlus*, vol. 5(647), 2016.

[25] Y. Song, H. Wang and T. Soyata, Hardware and software aspects of VM-based mobile-cloud offloading, in *Enabling Real-Time Mobile Cloud Computing through Emerging Technologies*, T. Soyata (Ed.), IGI Global, Hershey, Pennsylvania, pp. 247–271, 2015.

[26] J. Szefer, E. Keller, R. Lee and J. Rexford, Eliminating the hypervisor attack surface for a more secure cloud, *Proceedings of the Eighteenth ACM Conference on Computer and Communications Security*, pp. 401–412, 2011.

[27] A. Thongthua and S. Ngamsuriyaroj, Assessment of hypervisor vulnerabilities, *Proceedings of the International Conference on Cloud Computing Research and Innovations*, pp. 71–77, 2016.

[28] R. Uhlig, G. Neiger, D. Rodgers, A. Santoni, F. Martins, A. Anderson, S. Bennett, A. Kagi, F. Leung and L. Smith, Intel virtualization technology, *IEEE Computer*, vol. 38(5), pp. 48–56, 2005.

[29] G. Wang, Z. Estrada, C. Pham, Z. Kalbarczyk and R. Iyer, Hypervisor introspection: A technique for evading passive virtual machine monitoring, *Proceedings of the Ninth USENIX Workshop on Offensive Technologies*, 2015.

[30] Xen Project, x86 Paravirtualized Memory Management (wiki.xen.org/wiki/X86_Paravirtualised_Memory_Management), 2019.

[31] Xen Project, Xen Project Software Overview (wiki.xen.org/wiki/Xen_Project_Software_Overview), 2019.

Printed in the United States
By Bookmasters

Printed in the United States
By Bookmasters